W9-CUK-505

BIOLOGICAL
PHYSICS

To learn more about AIP Conference Proceedings,
including the Conference Proceedings Series, please visit the webpage
http://proceedings.aip.org/proceedings

BIOLOGICAL PHYSICS

3rd Mexican Meeting on Mathematical and Experimental Physics

México City, México   10 – 14 September 2007

EDITORS
Leonardo Dagdug
Leopoldo García-Colín Scherer
Universidad Autónoma Metropolitana-Iztapalapa
México City, México

SPONSORING ORGANIZATIONS
El Colegio Nacional, México
The Universidad Autónoma Metropolitana, México
CONACyT, México
CINVESTAV-IPN, México

Melville, New York, 2008
AIP CONFERENCE PROCEEDINGS ■ VOLUME 978

PHYS

Editors

Leonardo Dagdug
Leopoldo García-Colín Scherer

Av San Rafael Atlixco # 186
México City
México

E-mail: dll@xanum.uam.mx
        lgcs@xanum.uam.mx

L.C. Catalog Card No. 2008920096
ISBN 978-0-7354-0497-7
ISSN 0094-243X
Printed in the United States of America

cm 4/23/08

CONTENTS

Preface

The Third Mexican Meeting on Mathematical and Experimental Physics was held at El Colegio Nacional in México City, México, from September 10 to 14, 2007. The event consisted of the Meeting Medal Lecture, delivered by Prof. Víctor Márquez, a public lecture by Eusebio Juaristi, five plenary talks by Christopher Cramer, Sergey M. Bezrukov, Francis Everitt, Edmund Bertschinger and Ranulfo Romo, and three parallel symposia namely, Cosmology and Gravitation, Biological Physics and Physical Chemistry. The response of the community was altogether enthusiastic with over 1,000 participants and around 80 speakers from all over the world, U.S.A, Europe, Japan, and Latin America. Each symposium consisted of invited talks divided in 10 plenary talks and 21 half-hour talks. The overall impact of the event was more than satisfactory.

The main objective of the Meeting continues with the philosophy adopted in the two previous events in 2001 and 2004 namely, to provide a scenario to Mexican advanced students and young researchers on frontier topics in both mathematical and experimental physics in order to keep them in contact with developments taking place in other parts of the world and at the same time, to motivate and support the younger generations of researchers in our country.This is the reason behind inviting as lectures some of the most distinguished experts in the subject of the conference. We hope that we will be able to continue celebrating this international gathering every three years.

The proceedings of this Third Meeting will consist of three volumes namely,

- Cosmology and Gravitation
- Biological Physics
- Physical Chemistry

We would like to express our gratitude to everyone who contributed to the success of this Meeting in particular to our invited speakers, national and foreigners. The impact of their contribution among our scientific public was outstanding.

The whole meeting and in particular the symposium of Biological Physics would not be realized without the financial support of El Colegio Nacional, CONACyT (Mexico), our institution the Universidad Autónoma Metropolitana (UAM) and the Center for Advanced Studies and Research (CINVESTAV, Mexico). We wish to thank Dr. José Lema L., Dr. Oscar Monroy H. and Dr. Verónica Medina B., General Rector, Rector of the Iztapalapa Campus, and Dean of the Division of Basic Science and Engineering of UAM, respectively, and to Dr. René Asomoza, Director of CINVESTAV for sponsoring this international and multidisciplinary event. We also want to thank the very efficient financial and secretarial support given by Ma. Eugenia López and Cristina Ortiz.

We specially thank all the staff of El Colegio Nacional not only for the warm and kind hospitality offered to all the participants but also for the impeccable logistic that prevailed during the event.

Leopoldo García-Colín S. and Leonardo Dagdug.

LEOPOLDO GARCÍA–COLÍN SCHERER

Leopoldo García–Colin Scherer was born in México City on 27th November, 1930. He received his B.Sc. in Chemistry from the National University of México (UNAM) in 1953. His thesis dealt with the thermodynamic properties of D_2 and HD as part of a project whose leader was one of Mexicos's foremost scientist, Alejandro Medina. He obtained his Ph.D. at the University of Maryland in 1959 under Elliott Montroll's supervision. There, he began his dedicated career as a teacher and researcher and started an important research program in the broad area of Statistical Mechanics, Kinetic Theory, Irreversible Thermodynamics, Critical Phenomena, Chemical Physics, and other related subjects. Prof. Garcia–Colin has been recognized as the founder of Statistical Physics research in Mexico.

Upon his return to México in 1960, he faced the very hard duty of constructing totally new research groups in those fields of sciences cultivated by him. This was an exhausting task, demanding time, effort, and an insight in the development of high level educative institutions and government laboratories with a solid scientific basis. The first task was to optimize the scientific formation of scientists, mixing the selection and preparation of students, with a program of graduate studies outside México in selected places abroad, including universities of the United States, England, Holland, Germany, Belgium, etc. The role of García–Colín was essential through his personal scientific relations, since through their disposition of accepting the supervision of post–graduate students they contributed to the enhancement of Statistical Mechanics in Mexico.

His research interests encompass many areas of Physics and Chemical–Physics, like statistical physics of non-equilibrium systems, non-linear irreversible thermodynamics, kinetic theory of gases and plasmas, phase transitions, non-linear hydrodynamics, and glass transition. He is author of more than 240 research papers, 60 on popular science, 17 books dealing with thermodynamics, statistical mechanics, non-equilibrium phenomena, and quantum mechanics. His textbook on Classical Thermodynamics is well known in Spanish speaking countries.

Along many years, besides the fields previously mentioned, his scientific interests have spurred important incursions on protein folding and transport theory in Astrophysics and Cosmology, Air Pollution Problems, Science Policy, and Educational Research. He has maintained a continuous correspondence with many scientists and travels frequently all over the world. He is an active member of various prestigious International Societies, like the Third World Academy of Sciences, for instance, since 1988.

He was founder and Professor of Physics of the Escuela Superior de Física y Matemáticas at the National Polytechnical Institute (IPN) (1960–1963). Subsequently, he went to the Universidad Autónoma de Puebla (1964–1966), and then to the Sciences Faculty of the UNAM (1967–1984). He was researcher at the National Institute for Nuclear Research (ININ) (1966–1967), and thereafter he became Head of the Processes Basic Research Section at Mexican Petroleum Institute (1967–1974). In 1974, he started the Department of Physics and Chemistry at UAM–Iztapalapa as Head (1974–1978). Since 1988 he is National Researcher, level III, and now, additionally, National Emeritus Researcher and Distinguished Professor at UAM–Iztapalpa.

Among his extra–academic activities Professor García–Colín was vice–president and later President of the Mexican Physical Society. He is an active member of the American Physical Society, of the American Association for Physics Teachers, of the Mexican Academy of Sciences, of the Mexican Chemical Society, and of the American Association for the Advancement of Science, among others. His work has been recognized through several honors and awards, among them the Physics Award from the University of Maryland (1956–1957), the Prize in Exact Sciences of awarded by the now Mexican Academy of Sciences (1965), and the Merit Medal from the Universidad Autónoma de Puebla (1965). He held the van der Waals Chair at the University of Amsterdam, Holland (1976), and received the National Prize of Sciences and Arts from the mexican government (1988), to mention the significant ones.

Professor Leopoldo García–Colín became member of El Colegio Nacional in 1977. His opening talk: Modern Ideas on Liquid–Gas Transition, was answered by Professor Marcos Moshinsky.

His scientific and teaching carrier has been also recognized by degrees of Doctor Honoris Causa by Universidad Iberoamericana (1991), by the University of Puebla (1995), and by the National University of México (2006). In 2007 Professor García–Colín became Emeritus Professor at UAM–Iztapalapa.

VÍCTOR MARQUEZ

Dr. Marquez's career at the National Cancer Institute started in 1971, shortly after obtaining his Ph.D. in Medicinal Chemistry, when he became a Visiting Fellow in the Medicinal Chemistry Section of the Developmental Therapeutics Program. While working under the supervision of Dr. John S. Driscoll, Dr. Marquez developed a novel hydantoin template for the delivery of nitrogen mustards specifically to the CNS. His research led to the design of the active brain antitumor agent, spiromustine. Phase I evaluation of spiromustine commenced in 1987, but neurotoxicity presented as alterations in cortical integrative functions led to its eventual withdrawal from further clinical use.

After a five-year period (1972-1976) Dr. Marquez worked as Director of Research in Cosmos Laboratories, Caracas, Venezuela. There, he developed a new orally active anaphylactic inhibitor (LC-6), which was patented under US pat. No. 4,167,577 on September 11, 1979.

In 1977, Dr. Marquez returned to the NCI as a Visiting Scientist and after receiving tenure in 1987, became Deputy Laboratory Chief of the Laboratory of Medicinal Chemistry and eventually Chief of the same Laboratory in 2000. During his career at the NCI, he has worked in drug design and synthesis, targeting important enzymes in cancer and viral diseases. His discoveries have helped translate elements of basic research into solutions for specific medical problems as demonstrated by the fact that two compounds, cyclopentenyl cytosine (CPE-C) and 9-(2',3'-dideoxy-2'-fluoro-beta-D-*threo*-pentofuranosyl)adenine, the latter also known as lodenosine, entered clinical trials in the late 1990s and early 2000 as antitumor and anti-AIDS drugs, respectively. Both drugs were later suspended from further clinical use due to unexpected toxicity.

Other important discoveries that have resulted from his work in the area of nucleosides and nucleotides have been based on mechanistic approaches of transition-state theory to inhibit key enzymatic reactions. Some of the most important highlights in this area are briefly summarized: (1) Design and synthesis of the most potent transition-state analogue inhibitors of the enzyme cytidine deaminase based on a novel 5-hydroxy-perhydro-1,3-diazepin-2-one nucleoside system. (2) Synthesis of thiazole-4-carboxamide adenine dinucleotide (TAD) and the corresponding metabolically stable phosphonate analogue (beta-methylene TAD), two of the most potent inhibitors of the enzyme inosine monophosphate dehydrogenase. (3) Synthesis and discovery of 3-deazaneplanocin A, the most potent inhibitor known against the enzyme S-adenosylhomocysteine hydrolase and the most effective compound against the Ebola virus. (4) Synthesis and discovery of cyclopentenyl cytosine (CPE-C), the most potent inhibitor known to date of the enzyme cytidine triphosphate synthase. (5) Design and synthesis of the first, acid-stable dideoxypurine analogue for the treatment of AIDS. The compound, 9-(2',3'-dideoxy-2'-fluoro-beta-D-*threo*-pentofuranosyl)adenine, also known as lodenosine is a powerful inhibitor of HIV reverse transcriptase. (6) Design and synthesis of conformationally locked carbocyclic nucleosides as potent antitumor and antiviral agents. The corresponding thymidine analogue (MCT) is the most potent pyrimidine inhibitor of herpes virus types 1, 2 and Kaposi sarcoma virus. (7) Design and synthesis of zebularine as an orally active DNA methylase inhibitor with antitumor activity. Zebularine was withdrawn from clinical consideration due to unexpected toxicity in monkeys. (8) Design and synthesis of oligodeoxynucleotides containing conformationally restricted abasic sites as the most potent inhibitors of cytosine DNA methyltransferase capable of reactivating tumor suppressor genes.

Concomitant to the work performed above, theoretical studies with conformationally locked nucleosides and oligonucleotides containing locked units have been published in a series of papers in the *Journal of the American Chemical Society*. These studies have helped understand the role of sugar conformation in determining the affinity of nucleosides and nucleotides for kinases and polymerases. These studies have also set the basis for the design of delayed polymerase chain terminators effective against the scission repair mechanisms of resistant HIV reverse transcriptase.

Finally, in a second line of research, in collaboration with Dr. Peter M. Blumberg, a series of designed diacylglycerol mimetics that bind in nanomolar concentrations to the regulatory domain of protein kinase C (PKC) have been discovered. Some compounds displayed isozyme selectivity for specific members of the PKC family and have shown to have potent antitumor activity in the NCI 60 cell line screen as well as exceptional apoptosis inducing properties.

To date, the entire portfolio of his research is contained in 288 publications, 11 book chapters and 30 patents. Dr. Marquez has received 100 invitations as a speaker to various symposia, universities and pharmaceutical companies.

HONORS AND AWARDS:

- U.S. Department of Commerce Inventor's award, 1979
- Medical Research Council Visiting Professor, Universities of Saskatchewan and Manitoba, Canada, 1984-85
- NIH Merit Award, 1992
- Gordon Conference on Purines, Pyrimidines and Related Compounds (Vice-Chair, 1993; Chair, 1995).
- National Cancer Institute, Division of Basic Sciences (DBS) Intramural Research Award, 1997.
- Pharmazie-Wissenschaftpreiss 2001 (Pharmazeutische Chemie). Sponsored by Phoenix Pharmahandel AG & Co KG, Mannheim, November 26 2001.
- Senior Biological Research Scientist (SBRS) appointment, January 2002
- Member of the Medicinal Chemistry Division Long Range Planning Committee. American Chemical Society, Division of Medicinal Chemistry. 2003-2005
- 2003 Belleau Memorial Lecturer, McGill Chemical Society, McGill University, Montreal, Canada. March 18, 2003.
- Intramural AIDS Targeted Antiviral Program (IATAP) award (co-investigator Dr. Stephen Hughes) 2005 and 2006.
- Federal Technology Transfer Act Cash Award, 1998.
- Federal Technology Transfer Act Cash Award, 1999.
- Federal Technology Transfer Act Cash Award, 2000.

- Federal Technology Transfer Act Cash Award, 2001
- Federal Technology Transfer Act Cash Award, 2002
- Federal Technology Transfer Act Cash Award, 2003
- Federal Technology Transfer Act Cash Award, 2004
- Federal Technology Transfer Act Cash Award, 2005
- Federal Technology Transfer Act Cash Award 2006

Translation Against An Applied Force.

Gary M. Skinner, Yeonee Seol & Koen Visscher

Dept. of Physics, University of Arizona, 1118 E. 4th Street. Tucson, AZ 85721, USA

Abstract. Ribosome structure and mechanism are largely conserved among all known forms of life. Therefore, the motions associated with translation may be among the most ancient and fundamental in biology. However, the molecular mechanism of translocation, the coordinated movement of tRNAs and associated mRNA on the ribosome, has eluded scientists and remains obscure. Single–molecule experiments using optical tweezers and fluorescence microscope are starting to shed new light on these questions. For example, we have observed that moderate forces reverse direction of motion and ribosomes seem to slip backward into the 5' direction along a poly(U) message. Although the detailed molecular mechanism for ribosome slippage is not fully understood, these observations raise interesting biological questions about *e.g.* –1 frameshifting. Is the –1 frameshift essential for HIV–1 replication a result of tension in the message? Single–molecule experiments open the way towards quantitative modeling of ribosome motion and related phenomena such as –1 frameshifting.

Keywords: translation, RNA elasticity, single–molecule.

INTRODUCTION

Ribosomes are amongst the most complex molecular motors known bearing in mind their size, the complex arrangements of RNA and proteins that make up their structure, and the biochemical cycle that underlies their effective motion along messenger RNA (mRNA). For example, the prokaryotic ribosome is a large ~2.5 MDa structure composed of more than fifty proteins and 3 ribosomal RNA molecules which make up about half of the total mass. This structural intricacy is in keeping with its functional complexity. Translation involves three main phases: initiation, elongation, and termination including ribosome recycling. Although initiation and termination are interesting in their own right, it is during the elongation cycle that processive motion of the ribosome along mRNA occurs and where the molecular motor properties of the ribosome emerge. The elongation cycle itself is complex involving the selection and binding of transfer RNA (tRNA) in accordance with the mRNA code; translocation, *i.e.* the coordinated motion of tRNA and associated mRNA through the ribosome; and the formation of the peptide bond [1]. A simplified scheme of the translational elongation cycle, ignoring any intermediate states that exist, is depicted in Fig. 1. Prokaryotic ribosomes have 3 binding sites for transfer RNA (tRNA) partitioned between the small 30S and large 50S subunit: the Aminoacyl or A site; the Peptidyl or P site; and the Exit or E site. At the start of the cycle, peptidyl tRNA resides in the P site. Binding of aminoacyl tRNA occurs as a complex of tRNA•EF–Tu•GTP in the vacant A–site, catalyzed by the elongation factor Tu (EF–Tu) and GTP hydrolysis. Once aminoacyl tRNA has been bound in the A–site, the peptide bond is formed at the peptidyl transferase center (PTC) catalyzed by

CP978, *Biological Physics, 3rd Mexican Meeting on Mathematical and Experimental Physics*
edited by L. Dagdug and L. García-Colín Scherer
© 2008 American Institute of Physics 978-0-7354-0497-7/08/$23.00

ribosomal RNA [1]. To prepare for the next round of elongation, the A–site is vacated during the translocation step in which peptidyl tRNA (still in the A–site) and the deacylated tRNA (in the P–site) are moved to the P and E–site, respectively. mRNA associated with the tRNA via codon–anticodon interactions is carried along in the process, leading to the effective displacement of ribosomes towards the 3'–end, presumably in steps of one codon, *i.e.* 3 bases, at a time along the mRNA. This translocation is known to be catalyzed by the binding and hydrolysis of EF–G•GTP [2, 3]. A full biochemical cycle giving rise to ribosome motion thus requires at least two GTP hydrolysis events catalyzed by two GTPases, EF–Tu and EF–G.

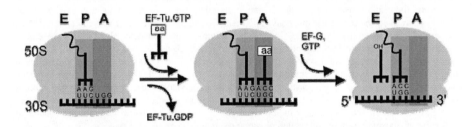

FIGURE 1. The translation elongational cycle. See Introduction.

Interestingly, translocation can also occur in the absence of EF–G, albeit extremely slowly [4], which argues that EF–G is unlikely to act as a motor protein that is physically pushing the ribosome along mRNA. Rather, the ribosome itself is the motor, and motion may have a strong diffusional component as there are no obvious conformational changes reminiscent of a power stroke seen for other types of motor proteins. The energy input required may be derived from a multitude of molecular interactions and processes within the ribosome, whereas directionality may simply result from the higher affinity of peptidyl tRNA for the P–site than for the A–site. However, because translocation requires the disruption of a number of contacts between the tRNAs and the ribosome, the energy barrier for forward motion can be expected to be quite considerable, which would explain the extremely slow elongation rate in the absence of EF–G. In such a simplified picture, the role of EF–G becomes clear as it may function to reshape the energy landscape to lower this energy barrier. Kinetic and biochemical analysis indicate that rapid hydrolysis upon binding of EF–G•GTP triggers conformational changes in both EF–G and the ribosome that "unlock" the ribosome after which tRNA–mRNA motion is thought to occur [5]. In particular, EF–G domain IV, which contacts the 30 S shoulder, is thought to play an essential role in opening up the decoding region. Deletion of this domain uncouples GTP hydrolysis and motion as shown by a ~1000–fold drop in the rate of elongation [1] while not affecting single round GTPase activity [2]. Interestingly, domain IV also mimics the anti–codon domain of the tRNA·EF–Tu complex and binds the A–site in the post translocation state [6], so that it may also serve to block any backward movement that might otherwise occur in the unlocked state. Interestingly, ribosomes have indeed been shown to spontaneously reverse movement in the absence of EF–G [7]. Such observations have begun to establish a picture of the ribosome as a thermal ratchet, where diffusion is biased towards mRNA's 3'–end in the unlocked state. This is an elegant and physically appealing qualitative picture without however, we note, any quantitative justification at this time. Although EF–G seems to couple GTP hydrolysis to motion, it remains to be seen if such coupling is fixed or if the number of GTP molecules hydrolyzed per translocation step varies depending on conditions, such as, for example, an opposing force. Does each EF–G•GTP hydrolysis cycle result in a one–codon sized translocation step, or are multiple hydrolysis cycles required? How does such coupling between biochemistry and mechanical motion depend upon force? Single–molecule experiments are capable of shedding light on such questions, as has been demonstrated in the case of the kinesin motor protein [8].

SINGLE–MOLECULE FORCE EXPERIMENTS

We have developed a single–molecule *in vitro* assay to record the motion of ribosome along mRNA against an applied force to start answering some of these questions. One such experimental geometry is depicted in Fig. 2, in which ribosomes from *E. coli* have been surface–immobilized using a streptavidin binding aptamer inserted into the 23S ribosomal RNA [9]. The cover–glass surface has been coated with a mixture of PEG and biotinylated PEG to facilitate specific binding of the ribosome while blocking any non–specific surface interactions. As messenger RNA we have used poly(U), which codes for polyphenylalanine, 3'–end labeled with dixoxygenin to attach a micron–sized bead that is held with the optical tweezers. We chose poly(U) as messenger RNA because it lacks any secondary structure, so that the ribosome displacement can readily be computed from observed bead displacements in the optical tweezers. Furthermore, in vitro translation of poly(U) has been rather well characterized by biochemical methods.

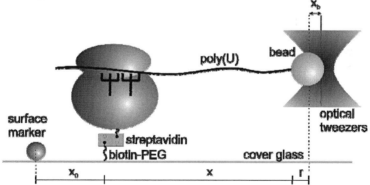

FIGURE 2. Single–molecule assay for recording ribosome motion along mRNA, against an applied force. For further information, see text.

The position of the bead held in the optical tweezers is monitored by evaluating the distribution of the forward scattered laser light with a quadrant photodiode positioned in the back focal plane of the microscope condenser [10, 11]. Upon addition of elongation buffer containing purified protein factors, synthetases, amino acids, tRNA,

3

GTP, ATP etc, motion of the ribosome along mRNA can be computed from the displacement of the bead being pulled out of the optical tweezers. The force exerted is computed by multiplying this displacement and the spring constant of the optical tweezers as determined by the equipartition theorem or from the power spectrum of the motion of the bead in the optical tweezers [12]. Rather then allowing the ribosome to pull the bead out of the optical tweezers, digital feedback keeps the bead at a fixed position within the optical tweezers by repositioning of the piezo–driven microscope stage. Thus, a constant force is maintained [8, 12], and ribosome displacements can be directly computed from the displacements of the stage. However, to eliminate adverse contributions to the signal due to specimen stage drift, video–based particle tracking has been used to determine the distance between the bead held in the force clamp and a surface immobilized marker ($x_0 + x + r$, see Fig. 2). Any change in this distance is directly proportional to the displacement of the ribosome along mRNA so long the force is maintained constant. The proportionality factor depends upon force according to the worm–like chain model that characterizes the elasticity of the mRNA. Following Moroz and Nelson expression for the relative extension [13]:

$$dx = \left[1 - \frac{1}{2}\left(\frac{L_p F}{kT} - \frac{1}{32} \right)^{-1/2} \right] dL_c$$

with dx the change in extension, dL_c the change in contour length, L_p the persistence length, k the Boltzmann constant, T temperature, and F the force. The change in contour length, dL_c reflects the ribosome's displacement along the mRNA, and can be computed provided the persistence length of the single–stranded mRNA is known. We have determined the elastic properties and persistence lengths of poly(U), and other homopolymeric messages such as poly(A) and poly(C) by stretching individual molecules between a microscope cover glass and a microscopic bead held with the optical tweezers.

At room temperature, poly(U) behaves as a random coil and its properties are well described by a worm–like chain model for polymer elasticity [14]. The persistence length of poly(U) was found to be ~0.9 nm at 500 mM Na^+ concentration, and increasing when lowering salt concentrations. Such behavior is to be expected due to the negatively charged phosphodiester backbone of RNA. Electrostatic repulsive interactions aid initial extension of the molecule (Fig. 3b, a lower force is required to extend the molecule at low Na^+ concentrations), which is reflected in an increased effective persistence length. In fact the scale–dependency of the effective persistence becomes very clear at 5 mM Na^+ concentration. Due to the electrostatic repulsion between the phosphate groups the effective persistence length is dependent upon extension: at small extension it is higher than at large extensions, by as much as a factor of 2–3 [14, 15]. Single stranded RNA and ssDNA are particularly good systems to study such behavior because the intrinsic persistence length, which is much smaller than that of dsDNA (~50 nm), is of the order of typical values of the Debye–Hückel screening length. In the case of dsDNA, the large intrinsic persistence length obscures any scale–dependency [14]. It will be of interest to further investigate these properties as a function of multi–valent ions that can also bind to RNA.

a
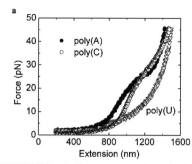

b
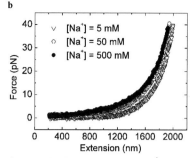

FIGURE 3. a) Force vs. extension of poly(U), poly(A) and poly(C) at 500 mM Na$^+$ concentration. **b)** Force vs. extension for poly(U) at varying Na$^+$ concentrations.

The force–extension curves for poly(A) and poly(C) differ markedly from those of poly(U) and show a distinct hump or plateau (Fig. 3). The presence of such a hump is generally indicative of a conformational change giving rise to an increase in the contour length of the molecule. Because both single–stranded poly(A) and poly(C) have been known to exist as helical structures at room temperature, we have interpreted the hump to reflect the helix–coil transition occurring upon stretching of the molecule [16]. These helices are thought to be stabilized by base–stacking interactions. We have fit our data to a theoretical model by Buhot and Halperin that predicts the elasticity of such single–stranded helical nucleic acids. This model assumes that helical domains predominantly are shorter than the typical persistence length of the helix, in which case this persistence length as used in the model can conveniently be chosen as infinite [17, 18]. When the random coil domains are assumed freely–jointed chains, an analytical solution of the force vs. extension then follows. Upon fitting this model to our force vs. extension data, values of the persistence length of the random coil, free energy of the base–stacking interaction, and Zimm–Brag cooperativity parameters are within the realm of published values [16, 19–24]. Future experiments at varying temperatures will further test the applicability of the model by Buhot and Halperin. Furthermore, stretching studies as function of the concentration of monovalent ions will provide insight into the role of electrostatics in stabilization of the helix. The results discussed here not only enable us to compute ribosome displacement from bead motion in the described single–molecule assay, but also go to show that homopolymeric RNA molecules may serve as ideal model systems for studying the elastic properties of polyelectrolytes in general.

Once we had determined the elastic properties of poly(U), the speed of a ribosome along the poly(U) message is estimated in units of codons/s and printed directly below the data records shown in Fig. 4. When the speed assumes a negative value, the mRNA tether shortens indicating the ribosome is moving towards the 3'–end (translocation), whereas a positive value indicates lengthening of the tether. The top panel in Fig. 4 (same ribosome–mRNA complex as in bottom panel) serves as a control in which GTP has been omitted (with otherwise identical buffer conditions) so that no motion is expected. At most forces the speed as observed in the presence of GTP exceeds that of the control. At a low applied force of 2.3 pN, a speed ~3 codon/s has been observed, which bearing in mind that the measurement has been done at room temperature is within the expected range.

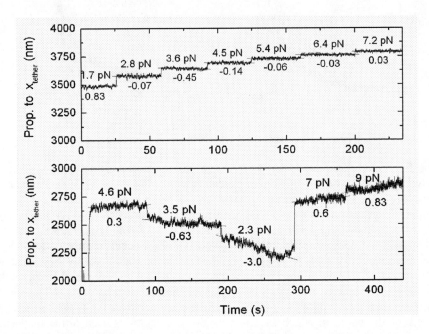

FIGURE 4. Change of mRNA tether length at varying set points of the force clamp. Upper panel is the control in the absence of GTP when motion is not expected, wheras the lower panel shows displacement data for the same ribosome mRNA complex in the presence of 1.5 mM GTP and 4.5 mM Mg^{2+}. Grey lines: linear fits to displacement data to obtain estimates of speed, with resulting speeds printed directly below the data record. Negative values indicate shortening of the mRNA tether.

At present, the experimental set up used lacks the spatial resolution to observe single codon–sized steps in these displacement data. At increasing forces, speeds decrease and in fact it appears that ribosome motion reverses so that the mRNA tether increases in length at forces of ~7.0 pN and higher. We have observed such backward slippage at rates as rapid ~1.3 codon/s at 7.5 pN of force (data not shown). Before discussing the possible molecular origin underlying the observed slippage, it is worth noting the increase in positional noise upon addition of GTP. At present, we have no explanation for this increase, but it is unlikely that it is due to small particulates that have entered the trap, because 1) we use a translation buffer reconstituted out of purified components, and 2) no such particulates were observed when turning off the tweezers. It is appealing to speculate about possible phenomena that may underlie such fluctuations. Is it a reflection of the proposed thermal ratchet behavior of this motor? It is worth noting that these data were taken using homopolymeric poly(U) as a message. Is the ribosome sliding back and forth along the message somehow? The latter seems unlikely because even a homopolymeric message has 5'–3' polarity that is likely to be sensed by the ribosome. Furthermore, work by Rachel Green`s lab, at zero force however, indicates that the length of the polypeptide synthesized is very tightly coupled to the length of the poly(U) message [25]. If the ribosome were to move back and forth along the poly(U) message, one would expect the polypeptides to have lengths exceeding those predicted for processive, unidirectional, codon by codon, decoding of the message. No such long polypeptides were observed and it was concluded that forward 3'–end directed motion is very strongly preferred [25]. However, one cannot rule out backward motion altogether as the probability per codon of any backward motion (at zero force) may just be too low to detect in the precipitation assay used, or rule out backward motion at opposing forces. Indeed, we have observed extended backward motion of ribosomes along poly(U), but only when forces of 7–9 pN opposing 3'–end motion have been applied. More questions than answers remain about this observation. Do polypeptides continue to be elongated during slippage, in which case one would expect the backward rate to be dependent upon concentrations of tRNA•EF–Tu•GTP? Or does force inhibit binding of tRNA•EF–Tu•GTP, or of EF–G? Konevega has demonstrated reverse motion of codon–anticodon associated tRNA and mRNA, a single codon worth, in the absence of EF–G [7]. The extent of backward displacement we have observed is larger than a single codon, and thus presumably would require disruption of the codon–anticodon interaction. Interestingly, there is precedent *in vivo* for such force–induced disruptions, which are thought to give rise to so–called –1 frameshifting.

Because the mRNA is encoded three bases—one codon—at a time, one mRNA molecule may in principle code for three different amino acid sequences. The reading frame of choice is defined at initiation of translation and maintained with high accuracy [26]. However, "programmed" frameshifts may occur in response to *cis*–acting elements that are generally found in mRNA of retroviruses [27, 28]. During a –1 frameshift, the codon–anticodon interactions are disrupted and tRNAs move relative to the mRNA so that the reading frame is displaced by one base towards the 5'–end of the mRNA. This reprogramming of translation generally leads to stop codon read–through and subsequent synthesis of a poly–protein, which upon further post–translational processing yields enzymatic and structural proteins required for viral replication [27, 28]. Two mRNA sequence elements are essential for –1 frameshifting: 1) a so–called "slippery sequence" at the ribosome coding region where the shift itself takes place. The slippery sequence has a typical base–sequence of X XXY YYZ, where XXY is read in the zero frame and XXX is read in the shifted –1 frame (X and Y can be the same base) [28]. The slippery sequence is generally rich in uracils, providing a low energetic barrier for the disruption of the codon–anticodon interaction. 2) A RNA frameshift signal, such as a pseudoknot (or even a simple hairpin structure in the case of HIV–1), positioned downstream of the slippery sequence near the entrance of the ribosome mRNA tunnel. Because the ribosome's mRNA tunnel is too narrow to accommodate double–stranded mRNA, the translation machinery needs to unwind mRNA structure during the elongation cycle [29]. –1 frameshifting is thought to occur when the downstream pseudoknot resists unfolding, whereupon tension is built up in the mRNA and the ribosome pauses and eventually slips backwards on the slippery sequence by a single nucleotide [30–32]. When in the translational elongation cycle does this –1 frameshifting take place? It has long been assumed that the frameshift occurs during translocation where the largest displacements of tRNA and associated mRNA take place [33, 34]. However, an alternative view has been proposed by Plant and colleagues, who hypothesized that the frameshift occurs during the accommodation step, when aminoacylated tRNA is bound at the A–site [35]. Upon binding of tRNA in the A–site, there is a concomitant motion of 9 Å of the tRNA and associated mRNA [36]. The extra tension built up due to this extra extension of the spacer is thought to induce the frameshift. A crude, on–the–back–of–the–envelope calculation considering the elasticity of a polymer in narrow cylindrical channel suggests that such a small increase in extension may result in a tangible increase in tension (of order 1 pN), so that the idea cannot be discarded purely on physical reasons [37, 38]. In fact, the model may have some special appeal because subsequent binding of EF–G could then trigger forward motion. The assumed ratchet function of EF–G then also blocks any further backward slippage of the mRNA, leaving no other choice than to unfold the pseudoknot. [39]. A more recent and elegant cryo–electron microscopy study however, shows a mammalian ribosome stalled, presumably pre–frameshifting, with the pseudoknot intact at the entrance to the mRNA tunnel, and eEF2 (EF–G in prokaryotes) occupying the A–site and interacting with the P–site tRNA [40]. The most straightforward interpretation of this structure is that −1 frameshifting has to occur during translocation and not during accommodation. The structure shows that the P–Site tRNA is distorted with the anticodon raised through bending of the tRNA towards the A–site which is occupied by the tip of eEF2 (EF–G in prokaryotes). The P–site tRNA is prevented from moving all the way back to the A–site by eEF2, which also blocks binding of aa–tRNA into the A–site. The bending of this tRNA is thought to be relaxed when the codon–anticodon interaction is disrupted and repaired in the −1 frame. Thus, −1 frameshifting depends upon a delicate balance of mRNA unwinding activity, and P–site tRNA and eEF2 interaction, mediated through tension in the mRNA [8]. This scenario does not readily explain why a repeated −1 frameshift would not occur upon the next translocation cycle? From simple energetic arguments, only considering the codon–anticodon interactions (a severe over–simplification ignoring the role of the ribosome), the mRNA in the coding region (A and P site) does not seem to have become dramatically more "unslippery" after the frameshift. It is possible that the frameshift may have alleviated a previous stereo–chemical mismatch of the pseudoknot with the ribosome's unwinding machinery enabling subsequent unfolding. Quantitative studies would enable modeling of −1 frameshifting, which may provide a physical understanding about why the −1 frameshift is only by a single base, or does not occur two times in a row. The displacement data of ribosomes against an applied force are first step toward such a goal. In particular, data with poly(U) as a message is relevant, because the slippery sequence of HIV–1 is U UUU UUA, almost identical to poly(U). With that in mind we have unfolded individual HIV–1 hairpins (spaced between two DNA–RNA hybrid handles, [41]) that trigger −1 frameshifting (data not shown). Pulling speeds were chosen identical to typical ribosome translocation rates, in which case predominantly reversible unfolding and refolding were observed. The HIV–1 hairpin was found to unfold at an average force of 12.9±0.9 pN, with an increase in contour length was 24±3 nucleotides (mean±sd). The latter observation indicates that only the upper stem is unfolded at this force, which is consistent with the unstable character of the lower stem. These data only allows for preliminary models to be developed, as unfolding of isolated hairpins does not fully account for the interactions between the ribosome and hairpin.

Single–molecule experiments aimed at understanding the role of mechanical force in translation of mRNA are still in their infancy. Yet, as shown by our work, they carry great potential to shed light on outstanding question about mechanochemical coupling, -1 frameshifting, *i.e.* the detailed mechanics and physics underlying the ribosome motor.

REFERENCES

1. R. Green and H. F. Noller, *Annu. Rev. Biochem.* **66**, 679–716 (1997).
2. M. V. Rodnina, A. Savelsbergh, V. I. Katunin and W. Wintermeyer, *Nature* **385**, 37–41 (1997).
3. V. I. Katunin,, A. Savelsbergh, M. V. Rodnina, and W. Wintermeyer, *Biochem.* **41**, 12806–12812 (2002).
4. N. V. Belitsina, G. Z. Tnalina and A. S. Spirin, *Biosystems* **15**, 233–241 (1982).
5. A. Savelsbergh, V. I. Katunin, D. Mohr, F. Peske, M. V. Rodnina and W. Wintermeyer, *Mol. Cell* **11**, 1517–1523 (2003).
6. J. Frank and R. K. Agrawal *Nature* **406**, 318–322 (2000).
7. A. L. Konevega, N. Fisher, Y. P. Semenkov, H. Stark, W. Wintermeyer and M. V. Rodnina *Nature Struct. Mol. Biol.* **14**, 318–324 (2007).
8. K. Visscher, M. J. Schnitzer and S. M. Block, *Nature* **400**, 184–189 (1999).
9. A. A. Leonov, P. V. Sergiev, A. A. Bogdanov, R. Brimacombe & O. A. Dontsova, *J. Biol. Chem.* **278**, 25664–25670 (2003).
10. K. Visscher, S. P. Gross and S. M. Block, *IEEE. J. Sel. Top. Quant. Elect.* **2**, 1066–1076 (1996).
11. F. Gittes and C. F. Schmidt, *Opt. Lett.* **23**, 7–9 (1998).
12. K. Visscher and S. M. Block, *Methods Enzymol.* **298**, 460–489 (1998).
13. J. D. Moroz and P. Nelson, *Proc. Natnl. Acad. Sci. U.S.A.* **94**, 14418–14422 (1997).
14. Y. Seol, G. M. Skinner and K. Visscher, *Phys. Rev. Lett.* **93**, 118102 (2004).
15. J. F Marko and E. D. Siggia, *Macromol.* **28**, 8759–8770 (1995).
16. Y. Seol, Y., G. M Skinner, K. Visscher, A. Buhot and A. Halperin, *Phys. Rev. Lett.* **98**, 158103 (2007)
17. A. Buhot and A. Halperin, *Macromolecules* **35**, 3238–3252 (2002).
18. A. Buhot and A. Halperin, *Phys. Rev. E* **70**, 020902(R) (2004).
19. W. Saenger, *Principles of Nucleic Acid Structure* (Springer–Verlag, New York, 1984).
20. M. S. Broido and D. R. Kearns, *J. Am. Chem. Soc.* **104**, 5207–5216 (1982).
21. E. G. Richards, C. P. Flessel and J. R. Fresco *Biopolymers* **1**, 431–446 (1963).
22. J. Brahms, A. M. Michelson and K. E. Vanholde *J. Mol. Biol.* **15**, 467 (1966).
23. T. G. Dewey and D. H. Turner, *Biochem.* **18**, 5757–5762 (1979).
24. T. G. Dewey and D. H. Turner, *Biochem.* **19**, 1681–1685 (1980).
25. D. R. Southworth, J. L. Brunelle and R. Green, *J. Mol. Biol.* **324**, 611–623 (2002).
26. C. G. Kurland, *Annu. Rev. Genet.* **26**, 29–50 (1992).
27. L. Balvay, M. Lopez Lastra, B. Sargueil, J.–L. Darlix and T. Ohlmann, *Nat. Rev. Microbiol.* **5**, 128–140 (2007).
28. P. J. Farabaugh, *Annu. Rev. Gen.* **30**, 507–528 (1996).
29. B. S. Schuwirth, M. A. Borovinskaya, C. W. Hau, W. Zhang, A. Vila–Sanjurjo, J. M. Holton and J. H. Cate, *Science* **310**,827–834 (2005).
30. J. D Lopinski, J. D. Dinman and J. A. Bruenn, *Mol. Cell. Biol.* **20**, 1095–1103 (2000).
31. P. Somogyi, A. J. Jenner, I. Brierley and S. C. Inglis, *Mol. Cell. Biol.* **13**, 6931–6940 (1993).
32. D. P. Giedroc, C. A. Theimer and P. L. Nixon, *J. Mol. Biol.* **298**, 167–185 (2000).
33. R. B. Weiss, D. M. Dunn, M. Shuh, J. F. Atkins, and R. F. Gesteland, *New Biol.* **1**, 159–169 (1989).
34. E. Yelverton, D. Lindsley, P. Yamauchi, and J. A. Gallant, *Mol. Microbiol.* **11**, 303–313 (1994).

35. E. P. Plant, K. L. Jakobs, J. W. Harger, A. Meskauskas, J. L. Jacobs, J. L. Baxter, A. N. Petrov and J. D. Dinman, *RNA* **9**, 168–174 (2003).
36. H. F. Noller, M. M. Yusupov, G. Z. Yusupova, A. Baucom, and J. H. Cate, *FEBS Lett.* **514:** 11–16 (2002).
37. J. O. Tegenfeldt, C. Prinz, H. Cao, S. Chou, W. W. Reisner, R. Riehn, Y. M. Wang, E. C. Cox, J. C. Sturm, P. Silberzan and R. H. Austin, *Proc. Natl. Acad. Sci. U.S.A.* **101**, 10979–10983 (2004).
38. P. G. de Gennes (1979) *Scaling Concepts in Polymer Physics* (Cornell Univ. Press, Ithaca, NY).
39. J. Frank and R. K. Agrawal, *Nature* **406**, 318–322 (2000).
40. O. Namy, S. J. Moran, D. L. Stuart, R. J. Gilbert and I. Brierley, *Nature* **11**, 244–247 (2006).
41. J. Liphardt, B. Onoa, S. B. Smith, I. Tinoco Jr. and C. Bustamante, *Science* **292**, 733–737 (2001).

Diffusion-Controlled Reaction With a Spot on the Side Wall of a Cylinder Membrane Channel

Alexander M. Berezkovskii

Mathematical and Statistical Computing Laboratory, Division for Computational Bioscience, Center for Information Technology, National Institutes of Health, Bethesda, Maryland 20892.
berezh@speck.niddk.nih.gov

and

Leonardo Dagdug

Departamento de Fisica, Universidad Autonoma Metropolitana-Iztapalapa, 09340, Mexico DF, Mexico.
dll@xanum.uam.mx

Abstract. We develop a theory of diffusion-controlled reactions with a spot located on the sidewall of a cylindrical membrane channel that connects two reservoirs containing diffusing particles which are trapped by the site at the first contact. An expression for the Laplace transform of the rate coefficient $k(t)$ is derived assuming that the size of the site is small compared to the channel radius. The expression is used to find the stationary value of the rate coefficient $k(\infty)$, as a function of the length and radius of the channel, the radius of the site and its position inside the channel as well as the particle diffusion constants in the bulk and in the channel. Our derivation is based on the one-dimensional description of the particle motion in the channel including a binding site. The validity of the approximate one-dimensional description of diffusion was checked by three-dimensional Brownian dynamics simulations. We found that the one-dimensional description works reasonably well when the size of the site does not exceed 0.2 of the channel radius.

Keywords: Diffusion, Membrane channel, binding site.
PACS: 87.15.R

1. INTRODUCTION

This article deals with binding of diffusing particles to a spot located on the sidewall of a cylindrical membrane channel as shown in Fig. 1. The main step of our study is mapping of the three-dimensional problem of the particle diffusion in the channel, onto a one-dimensional one. We suppose that the size of the site is small compared to the channel radius, and describe binding in terms of absorption by a δ-sink with a prescribed

CP978, *Biological Physics, 3rd Mexican Meeting on Mathematical and Experimental Physics*
edited by L. Dagdug and L. García-Colín Scherer

trapping rate which is function of both, the size of the site and the channel radius. This approximation is checked by three-dimensional Brownian dynamics simulation in Section 3 after we formulate the problem in the next section. Comparison shows that the estimate works reasonably well when the ratio of the size of the site to the channel radius is small enough.

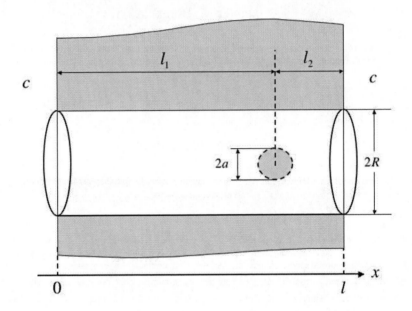

Figure 1. Circular binding site of radius a on the wall of the cylindrical channel of radius R. The channel connects two reservoirs separated by a membrane of thickness l. The reservoirs contain diffusing particles at concentration C. The site is located at distance l_1 from the left entrance into the channel and distance $l_2 = l - l_1$ from the right entrance.

We use this approximation in Section 4 to find the Laplace transform of the time-dependent rate coefficient, $k(t)$. This function characterizes survival probability, $S(t)$, of the site, which is assumed to annihilate at the first contact with the particle,

$$S(t) = \exp\left[-c\int_0^t k(t')dt'\right] \qquad (1.1)$$

where c is the particle concentration in the two reservoirs separated by the membrane. Our solution shows how the rate coefficient and, hence, $S(t)$ depend on the geometric parameters of the site and of the channel as well as on the location of the site inside the channel.

Our interest in this problem is motivated by recent progress in studies of transport through large membrane channels [1-3]. Compared with ion-selective channels of neurophysiology large channels seem to serve a different purpose than just to conduct small ions. These channels are rather pathways for metabolites and macromolecules such as proteins and nucleic acids, and their function is to regulate metabolite fluxes across the cell and organelle membranes. One of the new developments in studies of large channels is based on the idea that by measuring the current carried by small ions one can access the transport of larger molecules through their occlusion of the current [4]. These transient occlusions generate measurable excess noise and in many cases can be resolved as single-molecular events reporting on partitioning and dynamics of molecules within the confines of the channel pores.

One of the mechanisms that affect the occlusion of the small ion current is binding of the large solutes to their high affinity sites on the channel wall forming proteins. In the previous study [5] we analyzed the diffusion-controlled binding assuming that the site is large in the sense that its size is comparable with the channel radius. In this case a particle reaching the channel cross-section containing the site is trapped with probability close to unity. Therefore, this cross-section can be considered as perfectly absorbing boundary for the diffusing solute. However, recent experiments complemented by molecular dynamics simulations [6] have demonstrated that the sites can be relatively small, so that the particle reaching the cross-section of the channel where the site is located has a good chance to avoid contacting the site and to escape from the channel. A similar situation is realized for protons that protonate residues lining the channel pore [7].

With this in mind, in the present paper we extended the analysis given in Ref. [5] for large binding sites. Here we consider a model in which the cross-section containing the site is not necessarily perfectly absorbing. The trapping rate in this cross-section is now a free parameter of the model. The solution obtained in Ref. [5] is recovered from the solution derived below in the limiting case of infinitely high trapping rate, which corresponds to the perfectly absorbing cross-section. instructions.

2. FORMULATION OF THE PROBLEM

Consider two reservoirs separated by a membrane of thickness l. Reservoirs contain diffusing particles that can go from one reservoir to another through a cylindrical channel of radius R in the separating membrane. On the channel wall there is a circular binding site of radius a, $a \ll R$, which instantly loses its trapping ability (annihilate) when binding the first particle that reaches the site. This site is located at distances l_1 and $l_2 = l - l_1$ from the left and right entrances into the channel, respectively, as shown in Fig. 1. We assume that when the channel arises at $t = 0$ it is free from the particles which are uniformly distributed in the two reservoirs. The survival probability of the site given in Eq. (1.1) is the probability that no particle has reached the site by time t. Our goal is to find the rate coefficient $k(t)$ which is a function of the geometric parameters a, R, l, and l_1 as well as the diffusion constants of the particle in the channel, D_{ch}, and in the bulk, D_b.

To begin with, we note that the product $ck(t)$ is a time-dependent flux of the particles through the perfectly absorbing circular sink of radius a located on the channel wall exactly like the binding site. To find this flux one has to solve the three-dimensional diffusion problem in the two reservoirs and in the channel with absorbing boundary condition on the sink and to match the solutions at the channel entrances. This program is too complicated to be carried out. However, an approximate, but quite accurate technique has been found to handle such problems.[8] The idea is to describe three-dimensional diffusion in the channel as one-dimensional one with radiation boundary conditions at the channel ends that characterize the efficiency of escape from the channel of a particle approaching the channel boundary. Several problems, in which diffusion in the channel contacting with a bulk plays a crucial role, have been studied using this approach [5,8,9]. The results derived on the basis of the one-dimensional approximation were compared with those obtained in three-dimensional Brownian dynamics simulations. Excellent agreement was found between theoretically predicted and numerically obtained results.

Now we generalize the one-dimensional description of the particle motion in the channel so as to take trapping of diffusing particles by a small absorbing spot into account. Let the x-axis be directed along the channel and normal to the membrane, and $p(x,t)$ be the one-dimensional

14

density of the particles at point x of the channel, $0 < x < l$, at time t. We assume that this function satisfies the diffusion equation

$$\frac{\partial p}{\partial t} = D_{ch}\frac{\partial^2 p}{\partial x^2} - \kappa_a \delta(x - l_1)p, \qquad 0 < x < l \qquad (2.1)$$

in which the sink-term describes absorption of the particles by the spot. Taking advantage of the fact that the spot is small compared to the channel radius, $a << R$, we approximate the shape of the sink-term by the δ-function. The sink strength or trapping efficiency, κ_a, is taken to be equal to

$$\kappa_a = \frac{4D_{ch}a}{\pi R^2} \qquad (2.2)$$

In the next section we demonstrate that this description of trapping by a small circular absorber leads to the theoretical predictions which are in good agreement with the results found in three-dimensional Brownian dynamics simulations. We explain our choice of κ_a in Eq. (2.2) after we discuss boundary conditions imposed at the channel ends.

It has been shown in Ref. 8 that the end points $x = 0$ and $x = l$, as viewed from the channel, can be regarded as partially absorbing boundaries. The efficiency of escape from the channel of a particle that approaches the boundary is characterized by the trapping rate entering into the boundary conditions (BC), κ_{BC}, which is given by[8]

$$\kappa_{BC} = \frac{4D_b}{\pi R} \qquad (2.3)$$

To write the boundary conditions we introduce fluxes of particles that enter the channel from the left (L) and right (R) reservoirs at time t, $J_{L,R}(t)$. The boundary conditions can be written as

$$D_{ch}\frac{\partial p(x,t)}{\partial x}\bigg|_{x=0} = \kappa_{BC}p(0,t) + J_L(t)$$

$$(2.4)$$

$$-D_{ch}\frac{\partial p(x,t)}{\partial x}\bigg|_{x=l} = \kappa_{BC}p(l,t) + J_R(t)$$

15

It can be shown [8] that the boundary conditions more accurate than Eqs. (2.4) are non-Markovian, but reduces to Eqs. (2.4) on times larger than R^2 / D_b. In our further analysis we will neglect details of the kinetics occurring on such times. Then the fluxes $J_{L,R}(t)$ can be set equal to their stationary value, $4D_b Rc$, which is the stationary flux of the particles to a perfectly absorbing disk of radius R on the otherwise perfectly reflecting planar wall [10].

To explain our choice of the expression for the sink strength in Eq. (2.2) consider a particle diffusing in the channel with the absorbing spot on the wall assuming that the particle cannot escape from the channel because the exits are closed by reflecting lids. The average lifetime of such a particle, i.e., its mean first passage time to the spot is given by [11]

$$\tau_{3d} = \frac{V_{ch}}{4D_{ch}a} = \frac{\pi R^2 l}{4D_{ch}a} \qquad (2.5)$$

where $V_{ch} = \pi R^2 l$ is the volume of the channel and the subscript $3d$ indicates that this estimation of the lifetime is based on the three-dimensional consideration. We can also estimate this lifetime in the framework of the one-dimensional description. When the sink strength is small in the sense that the average lifetime of the particle is much greater than the average equilibration time which is of the order of l^2 / D_{ch}, the lifetime is given by

$$\tau_{1d} = \frac{l}{\kappa_a} \qquad (2.6)$$

where the subscript $1d$ indicates that this estimation is obtained in the framework of the one-dimensional description. One can see that our choice of κ_a in Eq. (2.2) leads to identity of the lifetimes in Eqs. (2.5) and (2.6), $\tau_{3d} = \tau_{1d}$.

In Section 4 we use the one-dimensional description of the particle diffusion and trapping in the channel to derive the Laplace transform of the rate coefficient, $k(t)$. However, first we discuss the numerical test of the approximate one-dimensional description in the following section.

3. NUMERICAL TEST OF THE APPROXIMATION

To check the accuracy of our approximate one-dimensional description of diffusion and binding in the channel we ran Brownian dynamics (BD) simulations in the geometry shown in Fig. 2 and compared the results found in simulations with those predicted on the basis of the one-dimensional description. In our simulations particles diffused in the cylindrical cavity of radius $R = 1$ and length l that contained a perfectly absorbing circular disk of radius a on its wall which was otherwise perfectly reflecting. The disk was located on equal distance from either end of the cylinder, which were also perfectly absorbing surfaces. Particles were initially uniformly distributed over the cross-section of the cylinder that passed through the center of the disk perpendicular to the wall. The particle trajectory was terminated at the moment when it reached the disk or the channel ends for the first time.

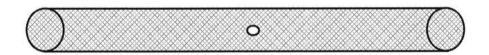

Figure 2. Cylindrical cavity of unit radius, $R = 1$, and length l containing a perfectly absorbing disk of radius a located on the cavity wall at equal distance from both ends of the cylinder, which are also perfectly absorbing surfaces, used in our Brownian dynamics simulations. The absorbing surfaces are cross-hatched in the figure. A special feature of our simulations is that near the absorbing surfaces we used 100 times smaller time step than in the rest of the cavity volume. The regions where the smaller time step was used are thin slabs of thickness 0.1 near the two absorbing ends of the cylinder and the rectangular parallelepiped of the height 0.1 and the square in the basis of the size $2(a + 0.1)$ surrounding the absorbing disk. The parallelepiped was located so that the center of its basis touched the wall of the cylinder just at the point where the center of the disk was located. In simulations the particle initial positions were uniformly distributed over the cross-section of the cylinder that passed through the center of the disk perpendicular to the cavity axis.

In simulations we ran $N = 50,000$ trajectories and recorded whether the trajectory crossed the disk or the channel ends and the lifetime, t_i for each trajectory, $i = 1,2,...,N$. We used these data to find the fraction of the trajectories trapped by the disk, $f_{disk}^{(BD)}$, and the average lifetime of the particles, $\bar{t}_{BD}$. This was done for $l = 10, 20, 30$ and $a = 0.05, 0.10, 0.15, 0.20$ assuming that the particle diffusion coefficient was equal to one half, $D = 1/2$. The results are shown in Tables 1 and 2. As might be expected, the probability that the particle is trapped by the disk increases with the disk size and with the length of the cavity. The average lifetime of the particle increases with the length of the cavity and decreases when the disk size increases.

	10	20	30
0.05	0.142	0.245	0.321
0.10	0.2471	0.4105	0.5008
0.15	0.343	0.515	0.616
0.20	0.409	0.591	0.682

Table1. The fraction of trajectories trapped by the disk, $f_{disk}^{(BD)}$, found in simulations for several values of the parameters l and a, which are given in the first row and the first column of the table, respectively.

	10	20	30
0.05	21.378	75.784	153.199
0.10	18.728	59.319	112.164
0.15	16.182	48.206	86.839
0.20	14.755	40.888	71.175

Table 2. The average lifetime of the particle, $\bar{t}_{BD}$, found in simulations for several values of the parameters l and a, which are shown in the first row and the first column of the table, respectively.

We also calculated these quantities in the framework of the one-dimensional description. The key function of this description is the particle propagator $G(x, t \mid l/2)$, which satisfies

$$\frac{\partial G}{\partial t} = D\frac{\partial^2 G}{\partial x^2} - \kappa\delta(x - l_1)G, \qquad 0 < x < l \qquad (3.1)$$

where D is the particle diffusion constant and the sink strength κ is given by Eq. (2.2) in which D_{ch} is replaced by D. The propagator also satisfies the initial condition $G(x, 0 \mid l/2) = \delta(x - l/2)$, and absorbing boundary conditions at $x = 0$ and $x = l$, $G(0, t \mid l/2) = G(l, t \mid l/2) = 0$. We solved this problem by the Laplace transformation method. Using the standard notation, $\hat{f}(s)$, for the Laplace transform of the function $f(t)$,

$\hat{f}(s) = \int_0^\infty \exp(-st) f(t) dt$, we can write the solution for the Laplace transform of the propagator as

$$\hat{G}(x,s\,|\,l/2) = A \times \begin{cases} \sinh\left(x\sqrt{\dfrac{s}{D}}\right), & 0 < x < \dfrac{l}{2} \\[2ex] \sinh\left((l-x)\sqrt{\dfrac{s}{D}}\right), & \dfrac{l}{2} < x < l \end{cases} \qquad (3.2)$$

where the factor A is given by

$$A = \left[2\sqrt{sD} \cosh\left(\frac{l}{2}\sqrt{\frac{s}{D}}\right) + \kappa \sinh\left(\frac{l}{2}\sqrt{\frac{s}{D}}\right) \right]^{-1} \qquad (3.3)$$

The equivalent of $f_{disk}^{(BD)}$ is the fraction of the particle trajectories trapped by the sink, $f_{disk}^{(1-d)}$. This fraction is given by

$$f_{disk}^{(1-d)} = \kappa \int_0^\infty G(l/2,t\,|\,l/2) dt = \kappa \hat{G}(l/2,0\,|\,l/2) \qquad (3.4)$$

Using the Laplace transform in Eq. (3.2) and the relation $\kappa = 4Da/(\pi R^2)$ we obtain

$$f_{disk}^{(1-d)} = \frac{1}{1 + \dfrac{4D}{\kappa l}} = \frac{1}{1 + \dfrac{\pi R^2}{al}} \qquad (3.5)$$

Survival probability of the particle, $S_{1-d}(t)$, is given by

$$S_{1-d}(t) = \int_0^l G(x,t\,|\,l/2) dx \qquad (3.6)$$

We can find its Laplace transform using the transform in Eq. (3.2). The result is

$$\hat{S}_{1-d}(s) = 2A\sqrt{\frac{D}{s}}\left[\cosh\left(\frac{l}{2}\sqrt{\frac{s}{D}}\right)-1\right] \tag{3.7}$$

The average lifetime of the particle, $\bar{t}_{1-d}$, is related to the transform $\hat{S}_{1-d}(s)$ by the relation, $\bar{t}_{1-d} = \hat{S}_{1-d}(0)$, and is given by

$$\bar{t}_{1-d} = \frac{l^2}{8D\left(1+\dfrac{\kappa l}{4D}\right)} = \frac{l^2}{8D\left(1+\dfrac{al}{\pi R^2}\right)} \tag{3.8}$$

Expressions in Eqs. (3.5) and (3.8) were used to find the probability of trapping by the disk, $f_{disk}^{(1-d)}$, and the average lifetime, $\bar{t}_{1-d}$, predicted by the one-dimensional description for the same values of the parameters R, a, l, and D, which where used in simulations. The ratios of the theoretically predicted and numerically obtained values are given in Tables 3 and 4. One can see that deviations of these ratios from unity are less than 10% even when the disk size is as large as 0.2.

	10	20	30
0.05		0.98	1.01
0.10	0.98	0.95	0.98
0.15	0.94	0.95	0.96
0.20	0.95	0.95	0.96

Table 3. The ratio of the theoretically predicted probability of the particle trapping by the disk, $f_{disk}^{(1-d)}$ in Eq. (3.5), to the value of this probability found in simulations, $f_{disk}^{(BD)}$, for several values of the parameters l and a, which are given in the first row and the first column of the table, respectively.

	10	20	30
0.05	1.01	1.00	0.99
0.10	1.01	1.03	1.03
0.15	1.05	1.06	1.07
0.20	1.04	1.08	1.09

Table 4. The ratio of the theoretically predicted average lifetime of the particle, $\bar{t}_{1-d}$ in Eq. (3.8), to the value of this lifetime found in simulations, $\bar{t}_{BD}$, for several values of the parameters l and a, which are given in the first row and the first column of the table, respectively.

Another comparison of the prediction based on the one-dimensional description with the result found in three-dimensional Brownian dynamics simulations is presented in Fig. 3 which shows logarithm of the particle survival probability as a function of time for $a = 0.05$ and $l = 30$. The solid curve was obtained by inverting numerically the Laplace transform of $\hat{S}_{1-d}(s)$ in Eq. (3.7), while the crosses represent the survival probability $S_{BD}(t)$ found in simulations. One can see that the two survival probabilities are very close to one another on times smaller than 900 dimensionless units. The difference between $S_{BD}(t)$ and $S_{1-d}(t)$ on these times is smaller than 4%. On larger times the difference becomes significant because of the poor statistics. The point is that the survival probability at $t = 900$ is less than $2 \cdot 10^{-3}$. Therefore the number of trajectories with the lifetimes greater than 900 in our simulations was less than one hundred that is definitely not enough for finding a reliable value of the survival probability.

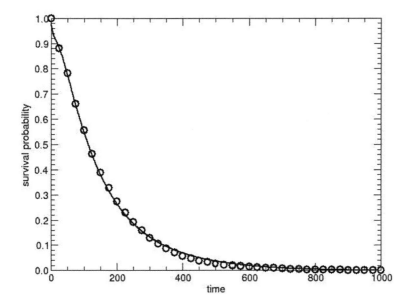

Figure 3. Particle survival probability in the cylindrical cavity of unit radius, $R = 1$, and length $l = 30$ containing a perfectly absorbing disk of radius $a = 0.05$ located on the cavity wall at equal distances from both ends of the cylinder, which are also perfectly absorbing surfaces, as a function of the dimensionless time. Open circles represent the survival probability found in simulations while solid

curve is obtained by numerically inverting the Laplace transform in Eq. (3.7). In simulations the particle initial positions were uniformly distributed over the cross-section of the cylinder that passed through the center of the disk perpendicular to the cavity axis.

To summarize, comparison with three-dimensional Browian dynamics simulations discussed in this section shows that the approximate one-dimensional description of diffusion and binding in the channel works reasonably well when the binding site is not too large. This is important since this approximation allows one to convert unsolvable three-dimensional problems into one-dimensional ones which can be solved with relative ease.

4. RATE COEFFICIENT

In this section we derive an expression for the Laplace transform of the rate coefficient, $k(t)$, which determines the survival probability of the binding site, Eq. (1.1). The derivation is based on the one-dimensional model formulated in Eqs.(2.2)–(2.4). It is assumed that when the channel is formed at $t = 0$, it is free from diffusing particles. Therefore, we will solve Eq. (2.1) with the initial condition $p(x,0) = 0$.

The one-dimensional density $p(x,t)$ can be written as a sum of two terms:

$$p(x,t) = p_L(x,t) + p_R(x,t) \qquad (4.1)$$

where $p_L(x,t)$ and $p_R(x,t)$ are due to the particles entering the channel through the left and right ends, respectively. These functions can be written in terms of the particle propagator in the channel, $G(x,t \mid x_0)$, which satisfies

$$\frac{\partial G}{\partial t} = D_{ch}\frac{\partial^2 G}{\partial x^2} - \kappa_a \delta(x - l_1)G, \qquad 0 < x < l \qquad (4.2)$$

with the initial condition $G(x,0 \mid x_0) = \delta(x - x_0)$ and the boundary conditions

$$\left[D_{ch}\frac{\partial G}{\partial x} - \kappa_{BC}G \right]\Bigg|_{x=0} = \left[D_{ch}\frac{\partial G}{\partial x} + \kappa_{BC}G \right]\Bigg|_{x=l} = 0 \qquad (4.3)$$

22

The expression for $p_L(x,t)$ and $p_R(x,t)$ in terms of the propagator and the entering fluxes are

$$p_L(x,t) = \int_0^t G(x,t-t'|0)J_L(t')dt' \qquad (4.4)$$

and

$$p_R(x,t) = \int_0^t G(x,t-t'|l)J_R(t')dt' \qquad (4.5)$$

To find the rate coefficient we need to know the flux through the sink, which is given by $\kappa_a p(l_1,t)$. Then the rate coefficient can be found by means of the relation

$$k(t) = \frac{1}{c}\kappa_a p(l_1,t) \qquad (4.6)$$

Using Eq. (4.1) we can write $k(t)$ as a sum

$$k(t) = k_L(t) + k_R(t) \qquad (4.7)$$

where

$$k_L(t) = \frac{\kappa_a}{c}p_L(l_1,t) = \frac{\kappa_a}{c}\int_0^t G(l_1,t-t'|0)J_L(t')dt' \qquad (4.8)$$

and

$$k_R(t) = \frac{\kappa_a}{c}p_R(l_1,t) = \frac{\kappa_a}{c}\int_0^t G(l_1,t-t'|l)J_R(t')dt' \qquad (4.9)$$

Replacing fluxes $J_L(t)$ and $J_R(t)$ by their stationary value $4D_b Rc$ we arrive at

$$k_L(t) = 4D_b R\kappa_a \int_0^t G(l_1,t-t'|0)dt' \qquad (4.10)$$

23

and

$$k_R(t) = 4D_b R\kappa_a \int_0^t G(l_1, t - t' \mid l) dt' \tag{4.11}$$

Below we derive the Laplace transforms of the rate coefficients $k_L(t)$ and $k_R(t)$ that are expressed in terms of the Laplace transforms of the propagators $G(l_1, t \mid 0)$ and $G(l_1, t \mid l)$ derived in Appendix A. As shown in this appendix these transforms can be written in terms of the Laplace transforms of the propagator $G(l_1, t \mid l_1)$ and the two fluxes $j_\infty(l_1, t)$ and $j_\infty(l_2, t)$. Eventually the Laplace transforms of the rate coefficients $k_L(t)$ and $k_R(t)$ are given by

$$\hat{k}_L(s) = \frac{4D_b R\kappa_a}{s} \hat{G}(l_1, s \mid 0) = \frac{4D_b R\kappa_a}{s} \hat{G}(l_1, s \mid l_1) \hat{j}_\infty(l_1, s) \tag{4.12}$$

and

$$\hat{k}_R(s) = \frac{4D_b R\kappa_a}{s} \hat{G}(l_1, s \mid l) = \frac{4D_b R\kappa_a}{s} \hat{G}(l_1, s \mid l_1) \hat{j}_\infty(l_2, s) \tag{4.13}$$

The Laplace transforms $\hat{G}(l_1, s \mid l_1)$ and $\hat{j}_\infty(x, s)$ are derived in Appendix A where we show that they can be written in terms of functions $P(x)$ and $Q(x)$ defined as

$$P(x) = \sqrt{sD_{ch}} \cosh\left(x\sqrt{\frac{s}{D_{ch}}} \right) + \kappa_{BC} \sinh\left(x\sqrt{\frac{s}{D_{ch}}} \right) \tag{4.14}$$

$$Q(x) = \sqrt{sD_{ch}} \sinh\left(x\sqrt{\frac{s}{D_{ch}}} \right) + \kappa_{BC} \cosh\left(x\sqrt{\frac{s}{D_{ch}}} \right) \tag{4.15}$$

The expressions are

$$\hat{j}_\infty(x, s) = \frac{\sqrt{sD_{ch}}}{P(x)} \tag{4.16}$$

and

$$\hat{G}(l_1,s\,|\,l_1) = [\kappa_a + \hat{j}_\infty(l_1,s)Q(l_1) + \hat{j}_\infty(l_2,s)Q(l_2)]^{-1} \qquad (4.17)$$

Eventually the Laplace transform of the rate coefficient in Eq. (4.7) is given by

$$\hat{k}(s) = \hat{k}_L(s) + \hat{k}_R(s) = \frac{4D_b R\kappa_a[\hat{j}_\infty(l_1,s) + \hat{j}_\infty(l_2,s)]}{s\left[\kappa_a + \hat{j}_\infty(l_1,s)Q(l_1) + \hat{j}_\infty(l_2,s)Q(l_2)\right]} \qquad (4.18)$$

This expression together with those in Eqs. (4.14)-(4.16) determines $\hat{k}(s)$ as a function of the geometric parameters a, R, l, and l_1, as well as the diffusion constants D_{ch} and D_b. This is one of the main results of the present paper. The Laplace transform in Eq. (4.18) is too complicated to be inverted analytically, but it can be easily inverted numerically. This allows one to find transient behavior of $k(t)$ from zero at $t = 0$ to its stationary value which is reached as $t \to \infty$, $k_{st} = k(\infty)$. The latter can be found from the small-s asymptotic behavior of $\hat{k}(s)$

$$k_{st} = \lim_{s \to 0} s\hat{k}(s) \qquad (4.19)$$

In the following subsection we use k_{st} found by means of this relation to analyze how the stationary rate coefficient depends on the geometric and kinetic parameters of the problem.

4.1. Stationary rate coefficient

Using the results of the first part of this section and the relations in Eqs. (2.2) and (2.3) we obtain

$$k_{st} = \frac{4D_b R(2+\lambda)}{(1+v\lambda)[1+(1-v)\lambda] + \dfrac{\pi R^2}{4al}\lambda(2+\lambda)} \qquad (4.20)$$

where

25

$$\lambda = \frac{4D_b l}{\pi D_{ch} R} \qquad \qquad v = \frac{l_1}{l} \qquad \qquad (4.21)$$

correspondingly, $1 - v = l_2 / l$. This stationary rate coefficient can be written in the form that has a transparent interpretation

$$k_{st} = 4D_b R[W(l_1) + W(l_2)] \qquad (4.22)$$

Here $4D_b R$ is the Hill formula[10] for the stationary trapping rate by a perfectly absorbing circular disk of radius R (entrance into the channel) located on the otherwise perfectly reflecting planar wall, while $W(l_1)$ and $W(l_2)$ are the probabilities to be trapped by the absorbing spot on the channel wall for particles entering the channel through the left and right ends, respectively.

The trapping probabilities $W(l_1)$ and $W(l_2)$ are given by

$$W(l_1) = W_{c-s}(l_1)K \ , \qquad \qquad W(l_2) = W_{c-s}(l_2)K \qquad (4.23)$$

Here $W_{c-s}(x)$ is the trapping probability found assuming that the entire cross-section (c-s) of the channel perpendicular to its axis and located at distance x from its entrance is perfectly absorbing,

$$W_{c-s}(x) = \left[1 + \left(1 - \frac{x}{l}\right)\right]^{-1} \qquad (4.24)$$

The factor K in Eq. (4.23) is the probability to be trapped by the absorbing spot on the channel wall for a particle that starts from the cross-section passing through the center of the spot perpendicular to the channel axis; the particle starting position is uniformly distributed over the cross-section. The factor K is given by

$$K = \left[1 + \frac{\pi R^2}{4al}\lambda(2 + \lambda)W_{c-s}(l_1)W_{c-s}(l_2)\right]^{-1} \qquad (4.25)$$

Thus, the stationary rate coefficient in Eq. (4.22) can be written as

$$k_{st} = 4D_b R[W_{c-s}(l_1) + W_{c-s}(l_2)]K \tag{4.26}$$

Putting here $K = 1$ we recover k_{st} derived earlier in Ref. 5 assuming that the entire cross-section is perfectly absorbing.

Alternatively, the stationary rate coefficient in Eq. (4.20) can be written in the Collins-Kimball form[12]

$$k_{st} = \frac{(4D_{ch}a)k_{c-s}}{4D_{ch}a + k_{c-s}} \tag{4.27}$$

Here $4D_{ch}a$ is the Hill formula [10] for the small perfectly absorbing disk on the wall of the channel and k_{c-s} is a sum of two rate coefficients each of which has the Collins-Kimball form

$$k_{c-s} = k_{c-s}(l_1) + k_{c-s}(l_2) \tag{4.28}$$

where $k_{c-s}(x)$ is the stationary rate coefficient in situation when the entire cross-section located at distance x from the channel entrance is perfectly absorbing, given by [5]

$$k_{c-s}(x) = \frac{(4D_b R)(\pi R^2 D_{ch}/x)}{4D_b R + \pi R^2 D_{ch}/x} \tag{4.29}$$

Assuming that in Eq. (4.27) $4D_{ch}a \gg k_{c-s}$ we obtain $k_{st} = k_{c-s}$ and recover the expression for k_{st} derived in Ref. [5].

Expressions in Eqs. (4.22) and (4.27) can be used to write the stationary flux of the particles into the absorbing spot on the channel wall, f_{st}, in the two forms,

$$f_{st} = 4D_b R[W(l_1) + W(l_2)]c \tag{4.30}$$

and

$$f_{st} = \frac{(4D_{ch}a)k_{c-s}}{4D_{ch}a + k_{c-s}}c \tag{4.31}$$

which allow different interpretation. The flux in Eq. (4.30) is a sum of two fluxes each of which is a product of the stationary flux to the channel entrance, $4D_b Rc$, and the decreasing factor $W(l_i)$, $i = 1,2$, due to the fact that the binding site is hidden in the channel. The expression in Eq. (4.31) may be interpreted as the Hill formula for the stationary flux to the absorbing disk on the channel wall, $4D_{ch}ac_{eff}$, which contains the effective concentration, c_{eff}, of the particles

$$c_{eff} = \frac{k_{c-s}}{4D_{ch}a + k_{c-s}}c \qquad (4.32)$$

which is smaller than the particle concentration, c, in the reservoirs.

As might be expected, the stationary rate coefficient monotonically decreases, as the site moves into the channel, and reaches its minimum value when $l_1 = l_2 = l/2$, i.e., $v = 1/2$. One can see this from Eq. (4.20). The decrease of k_{st} may be characterized by the ratio $k_{st}|_{v=1/2}/k_{st}|_{v=0}$ which is given by

$$\frac{k_{st}|_{v=1/2}}{k_{st}|_{v=0}} = \frac{1 + \lambda + \dfrac{\pi R^2}{4al}\lambda(2+\lambda)}{1 + \lambda + \dfrac{1}{4}\lambda^2 + \dfrac{\pi R^2}{4al}\lambda(2+\lambda)} \qquad (4.33)$$

the ratio is small and the decrease is significant when $\lambda \gg 1$. This may be due to either because $D_{ch} \ll D_b$ or because of $L \gg R$, and, of course, both factors may contribute.

5. CONCLUDING REMARKS

The present paper is devoted to kinetics of diffusion-controlled binding to a small circular site located on the wall of a cylindrical membrane channel. One of the main results of our analysis is the expression for the Laplace transform of the rate coefficient, Eq. (4.18), which determines the survival probability of the binding site, Eq. (1.1). We use this expression to find the stationary value of the rate coefficient reached as $t \to \infty$. Different representations of this quantity are given in Eqs. (4.20), (4.22), (4.26), and (4.27). They show how the stationary trapping rate depends on the geometric and kinetic parameters of the system. These parameters are the length and radius of the channel, the radius of the site and the distances from its center to the channel ends, which determines location of the site inside the channel, as well as the diffusion constants of the particles in the channel and in the bulk outside the channel. Although the theory is developed assuming that the site radius is much smaller than the radius of the channel, it turns out that when the radius of the site formally tends to infinity we recover the result derived for large binding sites in Ref. [5].

Our analysis is based on an approximate one-dimensional description of the particle motion in the channel suggested in Ref. [8]. To derive the results we have extended the formalism so as to include trapping of the particles by the binding site into consideration. To check the accuracy of our approximate one-dimensional approach we compare theoretical predictions derived in the framework of the one-dimensional description with the results obtained in three-dimensional Brownian dynamics simulations. There is a good agreement between the predicted and simulated results when the radius of the site does not exceed 0.2 of the channel radius. This agreement may be considered as a justification of the approximate one-dimensional approach which has been used in Section 4 when finding the solution for the Laplace transform of the rate coefficient.

The fact that the site is hidden in the membrane channel leads to a decrease of the binding rate compared to the situation where the same site is exposed on the membrane surface. For the first time this effects was studied by Samson and Deutch[13] in their theory of diffusion-controlled reactions with buried active sites. Samson and Deutch analyzed enzyme kinetics in situation where the active site of the enzyme was a small spherical cap buried inside an inert sphere that represented the enzyme molecule. In Ref. [5] the result for the stationary trapping rate derived in that paper for large binding sites hidden in membrane channels is compared with the stationary trapping rate given by the Samson-Deutch theory [13] in the corresponding limiting case. A special feature of kinetics analyzed in the present paper is that the binding site is small compared to the channel radius. Therefore, the particle reaching the cross-section of the channel containing the site has a good chance to avoid being trapped and to escape from the channel.

Finally, we note that the results obtained in the present paper for circular binding sites can be easily generalized to the case of non-circular sites. This can be done by prescribing the non-circular site an effective radius, a_{eff}, given by $a_{eff} = \left[AP/(2\pi^2)\right]^{1/3}$, where A and P are the area and perimeter of the site. Justification for this prescription is given in Ref. [14].

ACKNOWLEDGMENTS

We are grateful to Sergey Bezrukov and Attila Szabo for numerous very helpful discussions of the problem and related issues. This study was supported by the Intramural Research Program of the NIH, Center for Information Technology. L.D. also thanks for partial support to CONACyT (by the grant 52305) and the Camachos Foundation.

APPENDIX A: LAPLACE TRANSFORMS OF THE PROPAGATORS $G(l_1,t\,|\,0)$ AND $G(l_1,t\,|\,l)$

The propagator $G(x,t\,|\,0)$ is the probability density of finding the particle at point x, $0 < x < l$, at time t conditional on that the particle enters the channel from the left reservoir at $t = 0$ and has not escaped from the channel for time t. This propagator satisfies Eq. (4.2) with the initial condition $G(x,0\,|\,0) = \delta(x)$ and the boundary conditions in Eq. (4.3). To find the Laplace transform of this propagator at $x = l_1$, $G(l_1,t\,|\,0)$, we first consider an auxiliary problem assuming that the sink strength κ_a is infinitely high. This implies that the particle is instantly trapped by the sink as soon as reaches the point $x = l_1$ for the first time. We denote the propagator for the auxiliary problem by $G_\infty(x,t\,|\,0)$. This propagator satisfies

$$\frac{\partial G_\infty}{\partial t} = D_{ch}\frac{\partial^2 G_\infty}{\partial x^2}, \qquad 0 < x < l_1 \qquad (A.1)$$

with the initial condition $G_\infty(x,0\,|\,0) = \delta(x)$ and the boundary conditions

$$\left(D_{ch}\frac{\partial G_\infty}{\partial x} - \kappa_{BC}G_\infty\right)\bigg|_{x=0} = G_\infty\big|_{x=l_1} = 0 \qquad (A.2)$$

Solving this problem we find that the Laplace transform of the propagator $G_\infty(x,t|0)$ is given by

$$\hat{G}_\infty(x,s|0) = \frac{\sinh\left((l_1-x)\sqrt{\dfrac{s}{D_{ch}}}\right)}{P(l_1)} \qquad (A.3)$$

where function $P(x)$ is defined in Eq. (4.14). The flux $j_\infty(l_1,t)$ is defined by

$$j_\infty(l_1,t) = -D_{ch}\left.\frac{\partial G_\infty(x,t|0)}{\partial x}\right|_{x=l_1} \qquad (A.4)$$

This flux is formed by those realizations of the particle trajectory that reach the point $x = l_1$ for the first time at time t. As follows from Eqs. (A.3) and (A.4) the Laplace transform of this flux, $\hat{j}_\infty(l_1,s)$, is given by the expression in Eq. (4.16) with $x = l_1$.

We use the flux $j_\infty(l_1,t)$ and the propagator $G(l_1,t|l_1)$ to write the propagator of interest, $G(l_1,t|0)$, in the convolution form

$$G(l_1,t|0) = \int_0^t G(l_1,t-t'|l_1)\,j_\infty(l_1,t')\,dt' \qquad (A.5)$$

The Laplace transform of this propagator is

$$\hat{G}(l_1,s|0) = \hat{G}(l_1,s|l_1)\hat{j}_\infty(l_1,s) \qquad (A.6)$$

Thus, to find $\hat{G}(l_1,s|0)$ we have to find the transform $\hat{G}(l_1,s|l_1)$. The propagator $G(x,t|l_1)$ satisfies Eq. (4.2), the initial condition $G(x,0|l_1) = \delta(x-l_1)$, and the boundary conditions in Eq. (4.3). Its Laplace transform, $\hat{G}(x,s|l_1)$, satisfies

$$D_{ch}\frac{d^2\hat{G}(x,s|l_1)}{dx^2} - s\hat{G}(x,s|l_1) = \delta(x-l_1)\left[\kappa_a\hat{G}(l_1,s|l_1) - 1\right] \qquad (A.7)$$

with corresponding boundary conditions at $x = 0$ and $x = l$, which are the Laplace transforms of the boundary conditions in Eq. (4.3). Solving the problem for $0 < x < l_1$ and $l_1 < x < l$ and then matching the two solutions at $x = l_1$ one can find the propagator, $\hat{G}(x, s \mid l_1)$. At the point $x = l_1$ this propagator, $\hat{G}(l_1, s \mid l_1)$, takes the form given in Eq. (4.16). Eventually, one can find the Laplace transform of interest, $\hat{G}(l_1, s \mid 0)$, by multiplying the transforms $\hat{G}(l_1, s \mid l_1)$ and $\hat{j}_\infty(l_1, s)$ as given in Eq. (A.6).

Following the same way one can find that the Laplace transform of the propagator $G(l_1, t \mid l)$, which characterizes the particle entering the channel from the right reservoir, is given by the expression in Eq. (A.6), in which the Laplace transform of the flux $j_\infty(l_1, t)$ should be replace by the transform of the flux $j_\infty(l_2, t)$. The result is

$$\hat{G}(l_1, s \mid l) = \hat{G}(l_1, s \mid l_1) \hat{j}_\infty(l_2, s) \qquad (A.8)$$

Note that the Laplace transform $\hat{j}_\infty(l_2, s)$ is given by the expression in Eq. (4.16) with $x = l_2$. The relations in Eqs. (A.6) and (A.8) are used in Eqs. (4.12) and (4.13) which give the Laplace transforms of the rate coefficients $k_L(t)$ and $k_R(t)$.

REFERENCES

1. H. Bayley and C. R. Martin, Chem. Reviews **100**, 2575 (2000).
2. S. M. Bezrukov, J. Membr. Biol. **174**, 1 (2000).
3. A. Meller, J. Phys. Condens. Matter. **15**:R581-R607 (2003).
4. S. M. Bezrukov, I. Vodyanoy, and V. A. Parsegian, Nature **370**, 279 (1994).
5. L. Dagdug, A. Berezhkovskii, S. M. Bezrukov, and G. H. Weiss, J. Chem. Phys. **118**, 2367 (2003).
6. C. Danelon, E. M. Nestorovich, M. Winterhalter, M. Ceccarelli, and S. M. Bezrukov, Biophys. J. **90**, 1617 (2006).
7. B. Prod'hom, D. Pietrobon, and P. Hess, Nature **329**, 243 (1987); P. Hess, Annu. Rev. Neurosci. **13**, 337 (1990); S. M. Bezrukov and J. J. Kasianowicz, Phys. Rev. Lett. **70**, 2352 (1993); J. J. Kasianowicz and S. M. Bezrukov, Biophys. J. **69**, 94 (1995); T. K. Rostovtseva, T.-T. Liu, M. Colombini, V. A. Parsegian, and S. M. Bezrukov, Proc. Natl. Acad. Sci. USA **97**, 7819 (2000); E. M. Nestorovich, T. K. Rostovtseva, and S. M. Bezrukov, Biophys. J. **85**, 3718 (2003).
8. S. M. Bezrukov, A. M. Berezhkovskii, M. A. Pustovoit, and A. Szabo, J. Chem. Phys. **113**, 8206 (2000).
9. L. Dagdug, A. M. Berezhkovskii, S. Y. Shvartsman, and G. H. Weiss, J. Chem. Phys. **119**, 12473 (2003); A. M. Berezhkovskii, M. A. Pustovoit, and S. M. Bezrukov, J. Chem. Phys. **116**, 6216 (2002); **116** 9952 (2002); **119**,3943 (2003); A. M. Berezhkovskii and A. V. Barzykin, Chem. Phys. Lett. **383**, 6 (2004).
10. T. H. Hill, Proc. Natl. Acad. Sci. U.S.A. **72**, 4918 (1975).

11. I. V. Grigoriev, Yu. A. Makhnovskii, A. M. Berezhkovskii, and V. Yu. Zitserman, J. Chem. Phys. **116**, 9574 (2002).
12. F. C. Collins and G. E. Kimball, J. Colloid. Sci. **4**, 425 (1949).
13. R. Samson and J. M. Deutch, J. Chem. Phys. **68**, 285 (1978).
14. O. K. Dudko, A. M. Berezhkovskii, and G. H. Weiss, J. Chem. Phys. **121**, 1562 (2004).

Nonlinear Analysis of Time Series in Genome-Wide Linkage Disequilibrium Data

Enrique Hernández-Lemus*, Jesús K. Estrada-Gil*, Irma Silva-Zolezzi**,
J. Carlos Fernández-López *, Alfredo Hidalgo-Miranda**, and Gerardo
Jiménez-Sánchez**

*Computational Genomics Department and ** Basic Research Department,*
Instituto Nacional de Medicina Genómica,
Periférico Sur No. 4124, Torre Zafiro II, Piso 6 Col. Ex Rancho de Anzaldo,
Álvaro Obregón México, D.F. C.P. 01900, México

Abstract. The statistical study of large scale genomic data has turned out to be a very important tool in population genetics. Quantitative methods are essential to understand and implement association studies in the biomedical and health sciences. Nevertheless, the characterization of recently admixed populations has been an elusive problem due to the presence of a number of complex phenomena. For example, linkage disequilibrium structures are thought to be more complex than their non-recently admixed population counterparts, presenting the so-called ancestry blocks, admixed regions that are not yet smoothed by the effect of genetic recombination. In order to distinguish characteristic features for various populations we have implemented several methods, some of them borrowed or adapted from the analysis of nonlinear time series in statistical physics and quantitative physiology. We calculate the main fractal dimensions (Kolmogorov's capacity, information dimension and correlation dimension, usually named, D0, D1 and D2). We also have made detrended fluctuation analysis and information based similarity index calculations for the probability distribution of correlations of linkage disequilibrium coefficient of six recently admixed (mestizo) populations within the Mexican Genome Diversity Project [1] and for the non-recently admixed populations in the International HapMap Project [2]. Nonlinear correlations showed up as a consequence of internal structure within the haplotype distributions. The analysis of these correlations as well as the scope and limitations of these procedures within the biomedical sciences are discussed.

Keywords: Nonlinear Analysis, Population Genomics, Information Theory.
PACS: 89.75.Fb, 82.39.Pj, 87.15.Cc

INTRODUCTION

The genetic information of a living organism is encoded on its DNA sequence. Due to mutations and other genetic variations, often a gene or a biologically relevant DNA sequence is present in more than one different form. Each one of these alternative forms is called an allele. In population genetics, allele frequencies show the genetic diversity of a species population or equivalently the richness of its gene pool. Allele frequency is defined as the relative frequency of alleles of a given sub-class A_i among all of the alleles A in the population [3]. It is generally believed that alleles distribute randomly, but this occurs only under certain assumptions, including the absence of selection. When these conditions apply, a locus is said to be in Hardy-

CP978, *Biological Physics, 3rd Mexican Meeting on Mathematical and Experimental Physics*
edited by L. Dagdug and L. García-Colín Scherer
© 2008 American Institute of Physics 978-0-7354-0497-7/08/$23.00

Weinberg equilibrium (HWE). The frequency of each allele is used, for example, in the identification of genes affecting disease. *Linkage disequilibrium* (LD) is a term frequently used in genetics for the non-random association between alleles at two or more loci. LD describes a situation in which some combinations of alleles or genetic markers occur more or less frequently in a population than would be expected from a random formation of haplotypes from alleles based on their frequencies [4].

Genetic association studies that identify genetic loci associated with disease phenotypes are benefited both in terms of cost and computational burden, by the minimization of SNPs [1] to be genotyped. The knowledge that a significant fraction of the human genome could be organized into a series of high LD regions mainly separated by short segments in very low LD [5,6,7] has led to the development of a number of algorithms that can be used to select informative markers for association studies [8]. Marker-selection algorithms are commonly based on the assumption that the whole set of sequence variants within a certain region with large values of background LD carries out redundant information and thus can be significantly reduced to a *selected* subset of SNPs, also known as tagSNPs. These markers can label neighboring markers or, in some cases, a set of common haplotypes within a so-called *LD block*. Nevertheless, the problem of characterizing the structure and even the existence of these LD Blocks has been argued [9], in this sense, a genome-wide study of the correlations between SNPs for several populations is desired. In this work we choose to study the time series associated with such correlations by means of non-linear analyses.

Recent analyses support the hypothesis that due to the random genetic drift in finite populations, regions of high LD and limited haplotypic diversity will be generated in regions of uniform recombination if the recombination rate is sufficiently low [10]. It seems possible that a large part of the genome could show smaller blocks of largely stochastic origin [11,12]. In those regions, there may be very different patterns of LD in different populations because of their different recent histories and demographies. Additional studies have also showed that "*...LD patterns can be highly variable among populations both across and within geographic regions...*" [9]. This is yet another reason to look up for selected LD pattern structures.

Classical measures of LD between two loci, based only on the joint distribution of alleles at these loci, present noisy patterns. A distance-based clustering algorithm, usually performs hierarchical partitioning of an ordered sequence of markers into disjoint and adjacent blocks with hierarchical structure [6]. The haplotype block structure has been considered as an appealing model for visualizing LD patterns, after a number of recent studies suggested that SNPs across a relatively large chromosomal region could be parsed into blocks of various lengths, but it is not clear up to date how to determine the boundaries of an LD block, and in such case what is the possible structure of them with respect to their lenghts and distributions [9]. In practice, LD

---

[1] SNPs or Single Nucleotide Polymorphisms are the most common type of genetic variation in the human genome. These markers have a fixed size of one nucleotide base and two alleles.

blocks have be loosely defined as sets of contiguous markers that exhibit low haplotype diversity within blocks; and strong LD within blocks and sharp decay of LD between blocks. However, the actual visualization of LD patterns throughout the genome relies heavily on the way the block structure is defined. At this point, no agreement has been reached upon a universal definition of blocks, and as a result, the block structures identified by different groups via different rules carry different features. In this sense, it is necessary to develop new mathematical and statistical techniques to recognize and reconstruct the highly complex structure that lies behind LD patterns.

LINKAGE DISEQUILIBRIUM AND HAPLOTYPE DISTRIBUTIONS

Linkage disequilibrium is caused by such non-adaptive processes as population structure, inbreeding, and stochastic effects. In population genetics, linkage disequilibrium is said to characterize the haplotype distribution at two or more loci. Causal association between different marker alleles related to mutations are highly related to the statistics of linkage disequilibrium [13, 14]. A natural way to measure the deviation from linkage equilibrium is to compare the observed and expected genotype frequencies. The LD between marker SNPs i and j could be quantified simply by:

$$D_{ij} = f_{ij} - f_i f_j \tag{1}$$

Of course for two given SNPs we expect the LD to be symmetric (antisymmetric, indeed) so there is only one meaningful coefficient of linkage disequilibrium (the positive one). It is easy to note that D_{ij} quantification it is very limited so other LD measures have been proposed, one of them is the binary LD density, ρ_{ij} defined as follows:

$$\rho_{ij} = \frac{D_{ij}}{\sqrt{f_i(1-f_i)f_j(i-f_j)}} \tag{2}$$

An additional issue is that, unless the allele frequencies at the two loci are the same one cannot achieve perfect correlation, in order to overcome this it has been useful to define a coefficient, D' in the following manner:

$$
\begin{aligned}
|D'| &= \frac{-D_{ij}}{min\left(f_i(1-f_i), f_j(1-f_j)\right)} D_{ij}; & D_{ij} < 0 \\
&= \frac{D_{ij}}{min\left(f_i f_j, (1-f_i)(1-f_j)\right)} D_{ij}; & D_{ij} > 0
\end{aligned}
\tag{3}
$$

In recent times has been suggested a statistic that is claimed to be the best for association studies, the delta statistic δ_{ij} defined as follows [15]:

$$\delta_{ij} = \frac{D_{ij}}{f_i f_{ij}^*} \qquad (4)$$

Here f_{ij}^* refers to the least frequent alleles product frequency. Equation 4 is restricted to positive values of D_{ij}. The presence of different statistics points out to the fact that there are serious weaknesses in all the statistics.

Both D_{ij} and ρ_{ij} have null expectation of the correlation coefficient, so it is convenient to consider the square of the correlation coefficient [15], usually called r^2 or Δ^2.

$$r^2 = \rho_{ij}^2 = \frac{D_{ij}^2}{f_i(1-f_i)f_j(i-f_j)} \qquad (5)$$

We have chosen to use the r^2-statistic as given by equation 5 because of its wide use and direct statistical interpretation in terms of correlation functions. Various other quantities related to LD have been developed. In the genetic literature the wording *two alleles are in LD* usually means to imply that $\delta \neq 0$. Contrariwise, linkage equilibrium, denotes the case $\delta = 0$. A similar definition in terms of r^2 values is not so easily attained, but it could be stated, on quantitative grounds, that $r^2 = 1$ implies total LD whereas $r^2 = 0$ represents total linkage equilibrium.

This set of quantities is usually enough to characterize the biological behavior of a few marker SNPs. Nevertheless, with the advent of high performance genotyping techniques one is confronted with very large datasets (between 10^5 and 10^6 SNPs for each individual sample, with usual sampling universes of 10^2 to 10^3 individuals) that interact in a combinatorial fashion. Needless to say, a large amount of biologically relevant information has to be processed and interpreted. The search for certain statistically significant *patterns of linkage disequilibrium* (LD patterns) in genomic regions common to chosen populations would be thus a worth-pursuing enterprise. In order to understand the complex functional relationship between patterns of linkage disequilibrium across *genome-wide* regions one needs to use mathematical and statistical tools developed for the analysis of highly nonlinear problems. Some of these problems have arisen in past within the context of chaotic dynamic systems, statistical mechanics and physiological signals. In this work we will adapt and apply some of these tools in order to perform comparison between different LD patterns commonly present in contemporary biomedical research.

NONLINEAR ANALYSIS

We are looking to understand several characteristic features of complexity in LD-patterns. To determine the range and structure of correlations between markers and the degree of *wholeness* of these patterns we will make a *detrended fluctuations analysis* (DFA) of the associated data series. By means of DFA one is able to determine the extent of correlation and the plausibility of an underlying fractal structure in their distribution in a given dataset. It is also desired to attain a statistical quantification of similitude/dissimilitude between those patterns in different genomic segments (e.g. between different populations), to construct such a quantity we will use the method of *information based similarity index* (IBS).

The degree of complexity and self-similarity of the patterns is given by its fractal dimensions. Here we will calculate the main fractal dimensions D_0, D_1 and D_2. D_0 or *Kolmogorov Capacity* gives us a measure of the density of phase-space points within a given fractal set, thus a higher value of D_0 from the preceding integer dimension will imply a higher degree of self-organization in the structure. D_1, *the information dimension* corresponds with the amount of information encoded in a given fractal, thus a higher value of D_1 is related with a more redundant data set. Finally the index D_2, also termed *correlation dimension* is related with the density of correlations between points (in this case genetic SNP markers) in such a way that D_2 measures the proportion of correlated points in the given series.

Detrended Fluctuation Analysis

It is a well known fact that a bounded time series could be mapped to a self-similar process by integration. Very frequently biological data is given in the form of highly non-stationary time series. Non stationarity reflects the fact that the mean, variance and higher moments and *time* correlation functions are no longer invariant under time translation. In such non-stationary processes the usual integration procedure will exaggerate the non-stationary character of the data. In order to overcome this limitations, Peng, et al [16] introduced a modified root mean square analysis of the underlying random walk that has been termed detrended fluctuation analysis (DFA). DFA presents some advantages over typical methods of time series characterization such as Hurst analysis and spectral decompositions. In one hand DFA allows the detection of self-similar patterns in the original series even if it's embedded in an apparent nonstationary frame. DFA also avoids the spurious detection of artificial self-simmilarity due to trending of the probability distribution function. DFA first integrates the time series as follows. If $\Gamma(i)$ is a given time series (with discrete time-steps i), then the k-integrated value $\Gamma(k)$ is given by:

$$\Gamma(k) = \sum_{i=1}^{k} \left(\Gamma(i) - \hat{\Gamma} \right) \qquad (6)$$

here $\hat{\Gamma}$ is the average value of $\Gamma(i)$ in the considered interval. $\Gamma(k)$ gives a mapping from a time series to a self similar process. In order to measure the characteristic scale for the integrated time series, the integrated time is divided into isometrical bins of length n. For every bin it is performed a least squares linear fitting of the data, we call this fitting function the *trend* in that bin. The calculated coordinate of the straight line is denoted by $\Gamma_n(k)$. In order to detrend $\Gamma(k)$ we need to substract the linear local trend $\Gamma_n(k)$. For a particular bin size n, the characteristic lengthscale for the fluctuations in the integrated and detrended series is given by:

$$\mathsf{F}(n) = \sqrt{\frac{1}{N} \sum_{k=1}^{N} \left(\Gamma(k) - \Gamma_n(k) \right)^2} \qquad (7)$$

In the DFA algorithm, the value of $\mathsf{F}(n)$ is calculated for all the involved time scales (bin sizes) in order to get a relationship between $\mathsf{F}(n)$ and the bin size n. It is expected that $\mathsf{F}(n)$ will increase with n. A linear relationship on a log-log plot (i.e. a power law) will imply self-similarity of the related fluctuations. The slope of this log-log plot determines the scaling exponent α.

Thus,

$$\mathsf{F}(n): \; n^{\alpha} \qquad (8)$$

The DFA method has proven useful in revealing the extent of long-range correlations in seemingly irregular time series [16,17]. A value of α greater than 0.5 and less or equal to 1.0 indicate the presence of persistent long-range power law correlations. The case $\alpha = 1.0$ is a special case that has raised a lot of of interest both from physicists and biologists for many years and it corresponds to $1/f$ noise [18]. The case when $0 < \alpha < 0.5$ denotes the presence of power-law anti-correlations, such that large values are more likely to be followed by small values and vice versa. If $\alpha > 1$, correlations are no longer of a power-law form, anyhow, $\alpha = 1.5$ indicates brown noise, the integration of white noise [16].

Information Based Similarity Index

In the nonlinear analysis of highly complex time series, special attention should be given to the discovery of hidden information within the repetitive appearance of certain basic patterns embedded in the given signals. In order to detect and somehow quantify such underlying structures several methods have been developed. Information-theoretical based analysis have been applied to the haplotype block analysis in the past [19], and other entropy-related studies are in course. A very promising approach is based on the consideration of the linguistic properties of the symbolic dynamics associated with the time series under consideration. This approach based on the quantification of the so called Information-Based Similarity Index (IBS) [20] initially developed to work out the complex structure generated by the human heart beat time series. Nevertheless, IBS has proved to be a very powerful tool in the comparison of the dynamics of highly nonlinear processes. In the particular case to be considered here, we study the time series associated with the linkage disequilibrium patterns between several populations, in order to capture the major differences and use them to discriminate before proceeding to more involved population genomics calculations.

Let us consider a time series $\Gamma(i) = \{\Gamma_1, \Gamma_2, \ldots, \Gamma_N\}$, it is possible to classify each pair of successive points into one of the following binary states I_n, if $(\Gamma_{n+1} - \Gamma_n) < 0$ then $I_n = 0$; in the other case $((\Gamma_{n+1} - \Gamma_n) > 0)$ $I_n = 1$. This procedure maps the N step real-valued time series $\Gamma(i)$ into an $N-1$ step binary-valued series $I(i)$. Once we have this series, it is possible to define a binary sequence of length m (called an m-bit word). Each of the m-bit words w_k represents a unique pattern of fluctuations in a given time series. For every unitary time-shift λ, the algorithm makes a different collection W_λ of m-bit words over the whole time series, $W_\lambda = \{w_1, w_2, \ldots, w_n\}_\lambda$. It is expected that the frequency of occurrence of these m-bit words will reflect somehow the underlying dynamics of the original (real-valued) time series.

It is in this point that one is to recall that in studies of natural languages it has been observed that authors have different *word usage*, i.e. characteristic preferences for the frequency of use of every word, this fact has been reflected on a statistical linguistic phenomena known as Zipf's law [21]. This statistical principle has been applied in a wide variety of systems from natural language [22] to DNA sequences [23,24] and even musical structures [25]. In order to apply this concept to symbolic sequences, one should consider the frequency of every m-bit word and then sort them in descending order by frequency of occurrence, in this way we are able to write down a probability distribution function in the *rank-frequency* representation (RF-PDF). This RF-PDF represents the statistical hierarchy of symbolic words of the original time series [20]. In the theory of stochastic processes, two given symbolic sequences (chains or strings) are said to be *statistically equivalent* if they give rise to similar (or even identical) probability distribution functions. Following the very same order of ideas, Yang and coworkers [20] defined a measure of similarity (akin to statistical equivalence) between two time series by plotting the rank number of every m-bit word in the first time series with the rank for the same m-bit word in the second time series. Obviously if the two RF-PDFs are statistically equivalent, then the scattered points will lie *almost surely* in the diagonal line. In this sense, the average deviation of these points from the diagonal (i.e. $\theta = 45^o$) is a good measure of the distance (or dissimilarity) between these two time series.

Of course since the time series are supposed to be finite, the m-bit words are not equally likely to appear. The method introduces the likelihood of each word by defining a weighted distance Δ_m between two given symbolic sequences σ_1 and σ_2 as follows:

$$\Delta_m(\sigma_1,\sigma_2) = \frac{1}{2^m - 1} \sum_{k=1}^{2^m} |R_1(w_k) - R_2(w_k)| F(w_k)$$

(9)

$F(w_k)$ is the normalized likelihood of the m-bit word k, weighted by its given Shannon entropy, i.e.:

$$F(w_k) = \frac{1}{Z}\left[-p_1(w_k)\log p_1(w_k) - p_2(w_k)\log p_2(w_k)\right]$$

(10)

in this case, $p_i(w_k)$ and $R_i(w_k)$ represent the probability and rank of a given word w_k in the i-th series. The normalization factor in equation 10 is the total Shannon's entropy of the ensemble and is calculated as $Z = \sum_k \left[-p_1(w_k)\log p_1(w_k) - p_2(w_k)\log p_2(w_k)\right]$.

$\Delta_m(\sigma_1,\sigma_2)$ is called the Information Based Similarity Index (IBS) between series σ_1, and σ_2. One notices that $\Delta_m(\sigma_1,\sigma_2)\in[0,1]; \forall\sigma_1,\sigma_2; \forall m$. In fact one is able to consider $\Delta_m(\sigma_1,\sigma_2)$ as a probability measure. In the situation in which $\lim\Delta_m(\sigma_1,\sigma_2)\to 1$ the series are absolutely dissimilar, whereas in the opposite case given by $\lim\Delta_m(\sigma_1,\sigma_2)\to 0$ the two series become identical (in the statistical sense).

Fractal dimensions: Kolmogorov Capacity, Information Dimension and Correlation Dimension

Complex objects are often characterized in terms of its generalized or fractal dimensions. Fractal dimension has been used (within a biological context) as measure of contour complexity and as a quantitative morphological measure of cellular complexity [26] and has also been mentioned that it could be used as a taxonomic character [27,28]. In connection with genetic complexity the seminal work of the group of H. E. Stanley [16, 23] in the fractal character of DNA sequences, established a new standard to be taken into account in all nonlinear studies of genome structure. In terms of nonlinear analysis of time series and specifically of chaotic dynamics in population biology it has been argued that a higher value of the minimum fractal dimension implies a smoother and more persistent population trend [29], which is consistent with the statement that fractal structures are more robust and resistant to change by virtue of their redundancy and regularity [30]. Fractal dimensions are thus important quantities for the characterization of complexity. Generalized fractal dimensions have been used as tools for data mining in large datasets [31]. By far, the most widely studied fractal dimensions are the Kolmogorov capacity, the information dimension and the correlation dimension, also known as D_0, D_1 and D_2 respectively, that we define as follows.

Let A be a nonempty set with a metric, and let $N(r, A)$ be the minimum number of open balls of radius r needed to cover A. Then the *Kolmogorov capacity*, also called fractal dimension of capacity D_0, is given by [32]:

$$D_0 = \limsup_{r \to 0} \frac{\log N(r, A)}{\log(\frac{1}{r})} \qquad (11)$$

Let us now define the *Hausdorff dimension* of a set A. Denote by η a covering of A by a countable number of subsets η_k of diameter $r_k \geq r$. Given a positive real β we define $m^\beta(A) = \lim_{r \to 0} m_r^\beta(A)$ where $m_r^\beta(A) = inf\{\sum_{k=1}^{\infty} (r_k)^\beta\}$. When this limit exists we call m^β the Hausdorff dimension of set β. The minimum of the Hausdorff dimensions of a set for which the probability measure $\rho(A) = 1$ is called the *information dimension* D_1. The information dimension is related with Shannon's information content (it is in fact, proportional to it) hence a larger value of D_1 signifies a greater information content. In order to define the correlation dimension of a set A, we will consider points $x_i \in A$. Then we will define the integral correlation function (ICF) of a set following the Procaccia-Grassberger algorithm [33] as follows:

$$C(\varepsilon) = \lim_{N \to \infty} \frac{1}{N^2} \sum_{i,j; \, i \neq j} \Theta(\varepsilon - | x_i - x_j |) \qquad (12)$$

here Θ is Heaviside's unitary step function. If the limit exists, then we define the correlation dimension D_2 by:

$$D_2 = \lim_{\varepsilon, \varepsilon' \to 0+} \left[\frac{\ln \dfrac{C(\varepsilon)}{C(\varepsilon')}}{\ln \dfrac{\varepsilon}{\varepsilon'}} \right] \qquad (13)$$

D_2 measures the probability that two points chosen at random will be within certain distance of each other, hence the name correlation dimension. Although for many practical purposes D_0, D_1 and D_2 present very similar values and trends, sometimes it results useful to look upon each one of them in search of a deeper understanding of the complexity of the underlying phenomena. It is not common, though, to take into account higher q-values for the generalized fractal dimensions.

Methods and Materials

Genotyping was carried out on 300 non-related, self-defined *Mestizo* (recently admixed) individuals from the Mexican states of Guanajuato (GUA), Guerrero (GUE), Sonora (SON), Veracruz (VER), Yucatan (YUC) and Zacatecas (ZAC). DNA for genotyping was purified with the QiampMaxi DNA blood kit. Genotyping of all the samples was done according to the Affymetrix 100K SNP array protocol. The average call rate in our samples, using the BRLMM algorithm was 99.45 %. Samples below that threshold were repeated. Only SNPs that were present in > 80 % of the samples with a Hardy-Weinberg equilibrium p value cutoff of 0.0001 and < 1 % of heterozygosity at the X chromosome were included in our calculations. A total of 110,356 SNPs passed our quality criteria requirements in all the populations analyzed [1]. The 100K chip phased haplotypes of the HapMap samples were obtained from the International HapMap Project's web page [2] and extracted from the Phase II data. Phasing of the Mexican samples was computed using fastPhase v 1.1.4 [34]. All further analysis was done using phased haplotypes. LD calculations were carried out using Haploview software [35] and special-purpose code. The calculations involved in the DFA scaling exponent calculation and the IBS determination were both implemented using computational methods included in the PhysioToolkit library [36] via the DFA and IBS packages respectively. The values of the indicators or fractal dimensions D_0, D_1 and D_2 could be calculated with a variety of computational algorithms and codes. In terms of computational costs given the large amount of data involved in this project we choose to implement our estimates using the FD3 program [37] that makes use of the algorithm from Liebovitch and Toth [38] and works within computational times of order : $O(N \log N)$ where N is the number of data lines (points) input. All the calculations were done on a 40 node sub-cluster of the supercomputing unit of INMEGEN. Each node was constituted by a 5 Gb RAM Dual Opteron 64 bit Processor @ 2.4 GHz running under Linux/GPFS/CSM-IBM OS protocols.

RESULTS AND DISCUSSION

DFA

We calculated the scaling exponent for the DFA of the time series associated with LD patterns in the two arms of chromosomes 1 to 6, in six mexican mestizo populations (originary from the states of Guanajuato, Guerrero, Sonora, Veracruz, Yucatán and Zacatecas; GUA, GUE, SON, VER, YUC and ZAC respectively) from the Mexican Genomic Diversity Project [1] and in the non-recently admixed populations (Central Europeans from Utah, Japanese from Tokyo and Chinese Han from Beijing put-together, and Yoruba from Ibadan; CEU, JPT+CHB and YRI respectively) of the international HapMap project [2]. Thus, we characterized 12 DFA fractal objects for each population. In all 12 cases the highest rank was occupied by a non-recently admixed population. Recently admixed populations came one time as second highest and three times as third highest. This is an important issue since as α increases (getting closer to unity), the extent of long range correlations is also increased. Thus, eventhough in general recently admixed genomes have larger values of LD and more LD signals, it seems that the genomes of non-recently admixed populations are slightly more long-range correlated on its LD patterns than those on their admixed counterparts (see Table 1).

Table 1												
Population	C1(l)	C1(s)	C2(l)	C2(s)	C3(l)	C3(s)	C4(l)	C4(s)	C5(l)	C5(s)	C6(l)	C6(s)
GUA	0.986	0.981	0.988	0.981	0.988	0.985	0.991	0.990	0.980	0.981	0.976	0.980
GUE	0.984	0.979	0.982	0.983	0.987	0.985	0.992	0.997	0.976	0.985	0.971	0.976
SON	0.980	0.979	0.988	0.981	0.980	0.983	0.989	0.988	0.988	0.986	0.983	0.976
VER	0.984	0.977	0.989	0.978	0.991	0.978	0.997	0.989	0.982	0.981	0.971	0.981
YUC	0.982	0.982	0.983	0.978	0.990	0.983	0.997	0.987	0.981	0.980	0.978	0.982
ZAC	0.984	0.983	0.981	0.981	0.988	0.984	0.975	0.989	0.986	0.976	0.975	0.981
CEU	0.986	0.985	0.998	0.988	0.992	0.990	0.993	0.998	0.990	0.983	0.986	0.980
JPT+CHB	1.010	1.001	1.020	1.010	1.029	1.012	1.012	1.013	1.001	1.013	1.013	1.010
YRI	0.988	0.987	0.989	0.992	0.992	0.988	0.991	1.001	0.995	0.988	0.979	0.998

Table 1. Nonlinear statistics: Scaling exponents (α_j) for the DFA of the time series associated to linkage disequilibrium correlations across genomic distances in chromosomes 1 to 6. First value (l) corresponds to long chromosomal arm and second value (s) corresponds to short chromosomal arm. The value $\alpha = 1$ implies $1/f$ statistics or very long-range correlations.

The values of α for all 108 time series (12 individual calculations for each of the two arms of chromosomes 1 to 6 in 9 populations) distributed in the interval from 0.97114529 to 1.02871448, with an average value of 0.98825505 and a standard deviation of 0.01078211. All of these values were very close to unity (see table 1). As we already said a value of α : 1 indicates the presence of long range correlations of the $1/f$ type. $1/f$ or *pink noise* correlation is a condition in which all the frequencies are represented in the correlation structure, hence all the time scales are present in the dynamics of the phenomena, in contrast white-noise statistics typical of random behaviour is given by $\alpha = 0.5$. In figure 1 we observe the contrast between long range of correlations present in the LD structure of Chromosome 1, long arm in the CEU population and a random-generated *LD Pattern*. To remark the fact that this value of α ; 1, representative of long range statistics is indeed, caused by the persistence of correlations generated by LD pattern structure, we performed an *ad-hoc* simulation for the long arm of Chromosome 1 in all the populations. We automatically *masked* high values of r^2 (we choose as a limit $r^2 > 0.1$) for genomic distances greater than 1 Mb by multiplying them by 0.01. We then performed DFA to determine the extent of correlations as measured by α. Table 2 shows the results of this validation procedure. As it could be seen the value of α dropped-down from a long-range correlation value

$(\alpha ; 1)$ to a value representing medium range of correlations ($\alpha \in [0.6, 0.7]$), but still well-above the value for random processes ($\alpha = 0.5$).

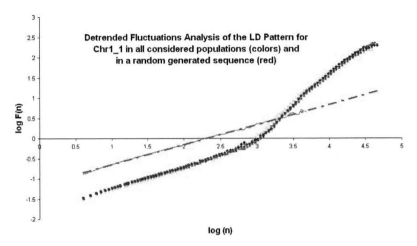

FIGURE 1. Detrended Fluctuations Analysis of the LD Pattern of Chromosome 1 long arm for all the considered populations, also a comparison with a completely random *LD Pattern*.

In figure 2, we show as an example the behaviour of an artificially de-correlated LD Pattern and the real LD Pattern for Chromosome 1, long arm in the CEU population. In figure 3 it is possible to see that this de-correlated pattern is midways between long ranged correlations and random behavior.

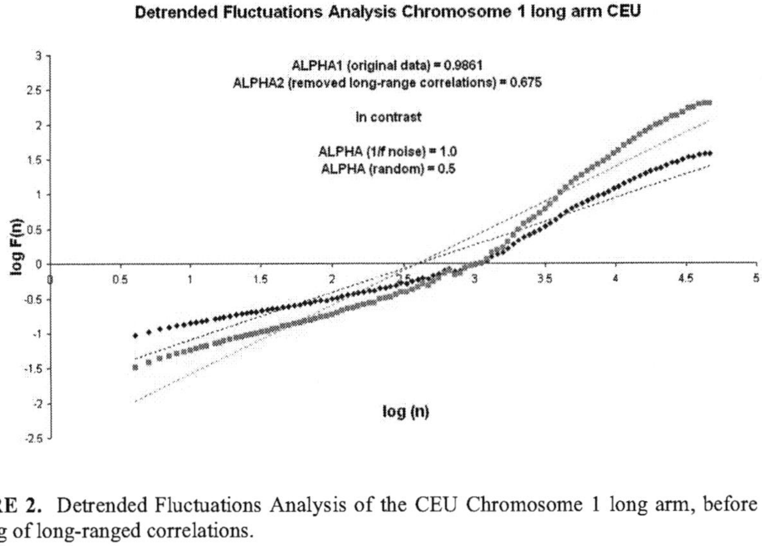

Detrended Fluctuations Analysis Chromosome 1 long arm CEU

ALPHA1 (original data) = 0.9861
ALPHA2 (removed long-range correlations) = 0.675

In contrast

ALPHA (1/f noise) = 1.0
ALPHA (random) = 0.5

FIGURE 2. Detrended Fluctuations Analysis of the CEU Chromosome 1 long arm, before and after masking of long-ranged correlations.

45

To further test for the relation between LD structure and long-range correlations we performed another simulation in which we decimated (i.e. we replace the series with a new series built by taking away, for example, 9 out of 10 equal-spaced points and letting the other points) the original r^2-time series for the long arm of chromosome 1 for the CEU population. We performed this decimation process with different moduli: 10, 20, 50, 100, 200, 500 and 1000, then making a DFA analysis for the decimated series. The results show that, in this case the LD pattern was a highly robust structure with regards to correlation persistence, since only after performing Mod1000 decimation (i.e. removal of 999 out of 1000 equally-spaced points) one obtains similar α-values to those of the modified series deprived of long range of correlations (see Table 3).

Table 2		
Population	C1(l)	C1(l) after removal of long-range correlations
GUA	0.986	0.673
GUE	0.984	0.685
SON	0.980	0.657
VER	0.984	0.607
YUC	0.982	0.657
ZAC	0.984	0.632
CEU	0.986	0.675
JPT+CHB	1.010	0.669
YRI	0.988	0.675

Table 2 Nonlinear statistics / synthetic de-correlation: Scaling exponents (α_j) for the DFA of the time series associated to linkage disequilibrium correlations across genomic distances in chromosome 1 long arm. The last column results from a simulation in which we masked correlations due to high values of r^2 for distances greater than 1 Mb

Table 3	
Chromosome 1, long arm, CEU	C1(l) after Mod_k decimation process
Original CEU	0.986
Mod10	0.977
Mod20	0.966
Mod50	0.963
Mod100	0.945
Mod200	0.873
Mod500	0.829
Mod1000	0.698

Table 3 Nonlinear statistics / decimation: Scaling exponents (α_k) for the DFA of the time series associated to linkage disequilibrium correlations across genomic distances in chromosome 1 long arm for the CEU population before and after Mod_k decimation

In the particular case treated here, the *time series* represents more of a *distance series*, for every data point comes from an equally spaced (in length across the genome) measurement of the correlation (as given by r^2-values) between two genetic markers. The afore-mentioned results show that all the genomic distances are represented in the correlation structure for every series. This means that long range correlations extend over the entire series, so that there is no *natural length-scale* associated with the LD structure (at least as measured by the r^2 coefficient) in the studied genomes. This is a striking result since some effort has been put in the past to determine the extent of LD-Blocks over a genome, i.e. as to what is the maximum distance in which LD is expected to show off between two genetic markers [39]. The present calculations show that even if very long range LD between 2 SNPs could be thought to be a rare event, it is not an impossible one; and in some cases it would be necessary to take into account the effect of the full LD structure within a genome.

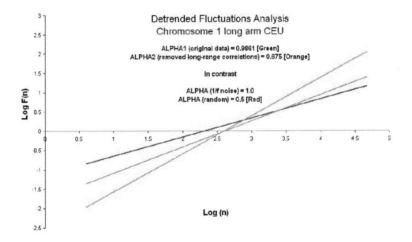

FIGURE 3. Detrended Fluctuations Analysis of the CEU Chromosome 1 long arm, before and after masking of long-ranged correlations, also a comparison with a completely random *LD Pattern.*

In a different context, the presence of slow-varying multiple-length variations in the power of *frequency* components in the form of 1/f noise has been observed to constitute a universal feature for DNA sequences [18]; this fact can also be translated to long-range of correlations. In these cases long-range order was seen to extend over distances of the order of 10^7 bases. It has been even proved that interspersed repeats are not responsible for this 1/f spectrum. Moreover, interspersed repeats have been demonstrated to affect predominantly the *high frequency* (i.e. small distance) spectrum. A word of caution was stressed however, the fact that the 1/f behavior is universal across different genomic regions do not imply that different regions exhibit the same *detailed* correlation structure but rather that they share the same governing principles [18]. This very same note should be given about the present results. The fact that all the considered genomes possessed a long-range behavior in their correlation structure does not imply, by any means, that their structure is identical.

IBS

For the considered comparisons (CEU versus All Other Populations, JPT+CHB versus All Other and YRI versus All Other) in chromosomes 1 to 6 we have sketched the surface plot of the IBS index distribution (Figure 4) for different values of the Bin Size, i.e. for different genomic distances (a value of k for the bin size represents a value of ($k+1$)-times 500 Kb). The minimum differences occur in the local regime (distances of the order of 1 Mb) where the functionality of the underlying genes make very similar (almost zero IBS index Δ_{min} on average) patterns in different populations, whereas the maximum differences correspond to wider genomic regions and are up to almost 27 % (Δ_{max} ; 0.27). The average difference across considered regions (maximum distance of about 10.5 Mb) is of the order of 7 % (Δ_{avg} ; 0.0697) with a very large mean standard deviation of Δ^{sd} ; 0.071. One can notice, from figure 4 that the IBS index Δ constitutes a good measure for comparison between populations up to a bin size k of 5 to 6 (i.e. genomic distances of 3 to 3.5 Mb) and a less good but still useful measure up to $k =$ 7 or 8 (4 to 4.5 Mb), since for greater distances all series are so divergent that one cannot have a well established discrimination criteria. Figure 5 shows an intricate yet somehow structured landscape in the distribution of bin-sized-averages of IBS between genomic regions across populations which may indicate some kind underlying goberning principle in the probability distribution of differences.

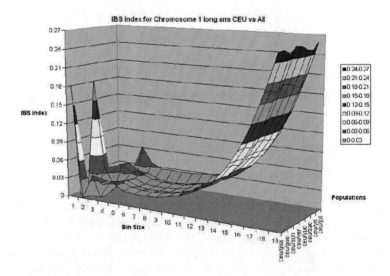

FIGURE 4. Full-Chromosome-Average IBS between CEU and all other studied populations for Chromosome 1 long arm.

To better establish the value and scope of the IBS index Δ, we performed a test-comparison between the original series for Chromosome 1, long arm in the CEU population (CEU-Chr1_1) with a) itself, b) a series absolutely proportional to CEU-Chr1_1, c) a random generated sequence of the same size and d) CEU-Chr1_1 vs GUE-Chr1_1. It could be seen in Figure 6, that both comparisons a) and b) showed Δ-values of zero for all bin sizes as expected. Also that comparison c) showed Δ-values in general different from zero and increasing with bin size. The same kind of trend was observed in comparison d) but as expected the differences are subtler than in comparison c). These results are also in strong concordance with the tendencies expected when considered both situations, hence Δ is a good index to quantify informative differences between time series in the present context.

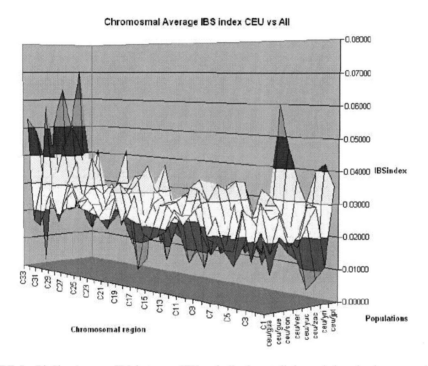

FIGURE 5. BinSize-Average IBS between CEU and all other studied populations in chromosomal regions.

Fractal dimensions D_0, D_1, D_2

The values of the fractal dimensions of chromosomes 1 to 6 for the considered populations ranged from 0.20696 to 0.36091, and were distributed as follows; the range of values for the Kolmogorov capacity D_0 is (0.20696 - 0.30593) with average $\overline{D_0} = 0.25770$ and standard deviation $D_0^{sd} = 0.01423$. The average density of points (in between that of single interspersed points ($D_0 = 0$) and a full line ($D_0 = 1$)) covers about 25 % of the phase space, indicating a medium degree of complexity in the landscape of correlations.

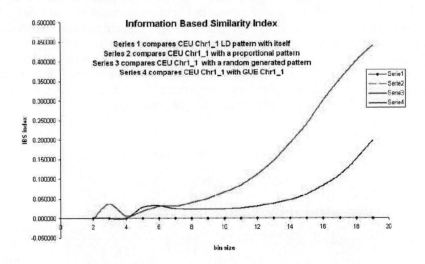

FIGURE 6. Information Based Similarity Index for CEU-Chr1_1 versus itself, a proportional pattern, a random pattern and GUE-Chr1_1

Table 4						
Population	C1(l,s)	C2(l,s)	C3(l,s)	C4(l,s)	C5(l,s)	C6(l,s)
GUA	0.20696, 0.24471	0.24089, 0.26267	0.24754, 0.26836	0.28173, 0.25941	0.26082, 0.24739	0.27923, 0.24511
GUE	0.24912, 0.24999	0.24213, 0.26645	0.24762, 0.26839	0.28673, 0.26119	0.26535, 0.25216	0.27832, 0.26330
SON	0.24098, 0.24401	0.23703, 0.25936	0.24321, 0.26059	0.27344, 0.25100	0.26181, 0.24127	0.26592, 0.25180
VER	0.24955, 0.24944	0.24420, 0.26398	0.24912, 0.26791	0.28241, 0.25902	0.27107, 0.25134	0.27878, 0.26006
YUC	0.24780, 0.24717	0.24089, 0.26666	0.24514, 0.26589	0.28137, 0.25973	0.26104, 0.25970	0.27897, 0.25933
ZAC	0.24448, 0.24720	0.24081, 0.26189	0.24180, 0.26364	0.27818, 0.25137	0.26214, 0.24437	0.26900, 0.25835
CEU	0.24459, 0.24883	0.24089, 0.30593	0.24773, 0.26115	0.27778, 0.25371	0.27100, 0.24255	0.26957, 0.25181
JPT+CHB	0.25956, 0.25930	0.24999, 0.27071	0.25595, 0.27892	0.28861, 0.26658	0.27319, 0.26287	0.25890, 0.26822
YRI	0.24290, 0.24117	0.24490, 0.24979	0.24293, 0.26316	0.27252, 0.24311	0.26290, 0.24466	0.24338, 0.26236

Table 4. Capacity Dimension D_0: Values of the Kolmogorov capacity dimension for the sets associated to LD patterns in considered populations

In the case of the information dimension D_1 the values are within (0.27564 - 0.36091) with average $\overline{D_1}$ = 0.30238 and standard deviation D_1^{sd} = 0.01462. This indicates that as much as 30 % of the information content within the considered time series is redundant.

Table 5						
Population	C1(l,s)	C2(l,s)	C3(l,s)	C4(l,s)	C5(l,s)	C6(l,s)
GUA	0.27564, 0.28953	0.28393, 0.30507	0.29272, 0.31499	0.32844, 0.30033	0.30587, 0.29094	0.32340, 0.28944
GUE	0.29500, 0.29579	0.28595, 0.30949	0.29281, 0.31519	0.33505, 0.30354	0.31170, 0.29731	0.32252, 0.30661
SON	0.28359, 0.28794	0.28063, 0.30022	0.28718, 0.30625	0.31883, 0.29521	0.30750, 0.28398	0.31197, 0.29665
VER	0.29492, 0.29464	0.28908, 0.30745	0.29475, 0.31391	0.32834, 0.29962	0.31522, 0.29605	0.32303, 0.30110
YUC	0.29287, 0.29176	0.28397, 0.31029	0.29003, 0.31224	0.32741, 0.30053	0.30653, 0.30055	0.32269, 0.29999
ZAC	0.28904, 0.29167	0.28386, 0.30400	0.28487, 0.30853	0.32469, 0.29604	0.30743, 0.28841	0.31556, 0.29895
CEU	0.29008, 0.29441	0.28436, 0.36091	0.29274, 0.30711	0.32402, 0.29922	0.31532, 0.28621	0.31630, 0.29687
JPT+CHB	0.30597, 0.30466	0.29674, 0.31672	0.30160, 0.32441	0.33756, 0.31119	0.31892, 0.30664	0.30403, 0.31356
YRI	0.28648, 0.28335	0.28850, 0.29434	0.28557, 0.30647	0.31495, 0.28626	0.30578, 0.28827	0.28197, 0.30437

Table 5 Information Dimension D_1: Values of the Information dimension for the sets associated to LD patterns in considered populations.

The correlation dimension D_2 lies within the range is (0.27437 - 0.35915) with average $\overline{D_2}$ = 0.30033 and D_2^{sd} = 0.014297. In connection with D_2 these values talk about circa 30 % of the phase points (i.e. r^2 values) are tightly correlated. Eventhough one may think that the values for different populations and chromosomes are somewhat similar, a closer look make evident a complex *landscape* in the distribution of fractal dimensions as can be seen in figures 7-9.

Table 6						
Population	C1(l,s)	C2(l,s)	C3(l,s)	C4(l,s)	C5(l,s)	C6(l,s)
GUA	0.27437, 0.28756	0.28210, 0.30337	0.29102, 0.31345	0.32524, 0.29951	0.30198, 0.28981	0.32101, 0.28817
GUE	0.29340, 0.29403	0.28447, 0.30767	0.29116, 0.31373	0.33228, 0.30198	0.30838, 0.29671	0.31998, 0.30481
SON	0.28145, 0.28558	0.27854, 0.29935	0.28522, 0.30412	0.31523, 0.29417	0.30371, 0.28253	0.30997, 0.29584
VER	0.29295, 0.29259	0.28731, 0.30568	0.29295, 0.31224	0.32460, 0.29872	0.31179, 0.29517	0.32083, 0.30037
YUC	0.29105, 0.28978	0.28214, 0.30847	0.28806, 0.31044	0.32383, 0.29968	0.30295, 0.29975	0.31995, 0.29896
ZAC	0.28702, 0.28963	0.28199, 0.30238	0.28298, 0.30657	0.32076, 0.29513	0.30352, 0.28707	0.31395, 0.29798
CEU	0.28843,	0.28274,	0.29087,	0.31986,	0.31186,	0.31462,

	0.29262	0.35915	0.30493	0.29854	0.28504	0.29619
JPT+CHB	0.30455, 0.30267	0.29556, 0.31528	0.30009, 0.32250	0.33483, 0.30977	0.31484, 0.30541	0.30135, 0.31198
YRI	0.28381, 0.28088	0.28566, 0.29320	0.28249, 0.30316	0.30904, 0.28458	0.30005, 0.28655	0.27868, 0.30247

Table 6 Correlation Dimension D_2: Values of the Correlation dimension for the sets associated to LD patterns in considered populations

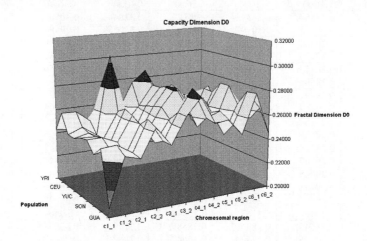

FIGURE 7. Capacity Dimension for the nine considered populations as a function of chromosomal region.

With regards to these findings we calculated the fractal dimensions in the artificially de-correlated series described above for chromosome 1 long arm for the CEU population. In the first case the Kolmogorov capacity diminished from 0.24459 to less than half this value ($D_0^{decorr} = 0.117$) indicating a lower density of phase points, hence a lower stochastic interaction within *phase-space*.

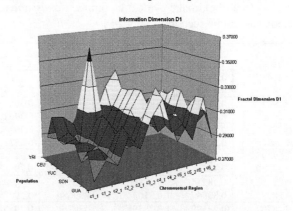

FIGURE 8. Information Dimension for the nine considered populations as a function of chromosomal region.

The information dimension also showed a significant change from a value of 0.29008 to $D_1^{decorr} = 0.1548$; that means that approximately 15 % of the information redundancy of this data series is lost after removal of long range correlations. Correlation dimension follows the same tendency, its value change from 0.28843 in the original CEU-Chr1_1 to $D_2^{decorr} = 0.1608$. It is worth mentioning that even if correlation dimension also diminished its value after synthetic de-correlation, it still implies about 16 % of strongly correlated points.

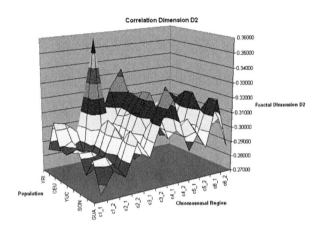

FIGURE 9. Correlation Dimension for the nine considered populations as a function of chromosomal region.

Concluding remarks

By considering the DFA results we have concluded that there is not a well defined LD-Block structure within the considered genomes, since the value of the scaling exponent (very near to 1) implies the existence of very long range correlations [16,18]. These results imply that is not completely improbable to find a high value of LD (at least, as measured by r^2-criteria) between two markers separated by a long genomic distance (even at the studied size of full-chromosomal dimensions). This means that at most one can consider these LD-Block structures as approximations to the real LD genomic structures, an observation that is consistent with the present understanding of LD pattern structures [9].

On the other hand, a careful analysis of the IBS indexes for the comparisons of the LD landscapes between populations show that examining the LD patterns could be a useful tool to distinguish between the genomes of different populations on an intermediate genomic length-scale between say 1 Mb and 3.5 Mb. For regions smaller than that, it could be confounding factors due to the physiological structure present in local segments of the genome, whereas for larger regions the LD patterns are so diverse that one cannot possible see a useful criteria of comparison. The fractal dimension analysis of the associated series showed that both non-recently admixed populations and recently admixed populations possess a highly complex (fractal) structure, characterized by dimensions ranging from approximately 0.2 to 0.36. A detailed analysis of the landscape associated with the fractal dimension distribution showed the somehow ubiquitous *autosimilar* phenomena characteristic of complex systems. A primary evaluation of the fractal dimensions of the considered LD-pattern time series, throws up very similar values in various populations and chromosomal segments but a more refined search shows an intricate landscape that tell us about the highly complex nature of the underlying genetic processes.

This kind of studies are expected to be useful as complimentary analyses in the search to understand and characterize LD structures, having in mind, both the biomedical implications (e.g. as tools for refining tagSNP searching algorithms or to compare the LD structures between different conditions in case control studies) and the basic biological principles underlying, for example, mechanisms of genetic recombination. Nevertheless one has to be cautious because it has been recently mentioned that long-range LD structures detected could be due to systematic errors, e.g. in the annotations for the genotyping chips or to undiscovered experimental draw-backs. Future studies relying on more refined technological platforms (e.g. denser SNPs sets per chip and improved anotations) will surely shed further light on these issues. Thus, after performing several analyses on a set of more than 1000 computational experiments over high-throughput genomic data, one final statement that arises is that a lot of work has to be done before we reach a complete understanding of the complex structures behind genome wide linkage disequilibrium patterns in both recently and non-recently admixed populations. Anyhow, the tools of nonlinear analysis have showed us some clues as to where to walk our steps.

REFERENCES

1. Hidalgo-Miranda, A., Estrada-Gil, J., Silva-Zolezzi, I., Uribe-Figueroa, L., Contreras, A., Balaam, E., Del Bosque-Plata, L., Barrientos, E., Lara-Álvarez, C., Goya-Ogarrio, R., Fernández-López, J.C., Hernández-Lemus, E., March, S., Dávila, C., Jiménez-Sánchez, G.; *The Mexican Genome Diversity Project: Analysis of genomic variability in Mexican Mestizo populations*, Manuscript in preparation, (2007).

2. The International HapMap Consortium. *A Haplotype Map of the Human Genome*, Nature 437, 1299-1320. (2005).

3. Gillespie, J.H., *Population Genetics A Concise Guide*, Second edition, The Johns Hopkins University Press, (2004)

4. Rife, D. C., *Populations of hybrid origin as source material for the detection of linkage Am. J. Hum. Genet.* **6**, p.26-33, (1954).

5. Daly, M.J., Rioux, J.D., Schaffner, S.F., Hudson, T.J., Lander, E.S., *High Resolution haplotype structure in the human genome*, Nat. Genet., 29, 229-232, (2001)

6. Patil, N., Berno, A.J., Hinds, D.A., *Blocks of limited haplotype diversity revealed by high resolution scanning of human chromosome 21*, Science 294, 1719-1723, (2001)

7. Gabriel, S.J.,Schaffner, S.F.,, Nguyen, H., et al, *The structure of haplotype blocks in the human genome, Science*, 296, 2225-2229, (2002)

8. Wang, N., Akey, J.M., Zhang, K., Chakraborty, R., Jin, L., *Distribution of recombination crossovers and the origin of haplotype blocks: the interplay of population history, recombination and mutation, Am. J. Hum. Genet.*, 71, 1227-1237, (2002)

9. Gu, S., Pakstis, A. J., Li, H., Speed, W. C., Kidd, J. R., Kidd, K. K., *Significant variation in haplotype block structure but conservation in tagSNP patterns among global populations*, Eur. Jou. Hum. Genet. 15, 3, 1-11, (2007)

10. Sawyer, S.L., Mukherjee, N., Pakstis, A.J., Feuk, L., Kidd, J.R., Brookes, A.J., and Kidd, K. K., *Linkage disequilibrium patterns vary substantially among populations*, Eur. Jou. Hum. Genet., 13, 677-686, (2005)

11 Chakraborty, R. and Weiss, K. M., *Admixture as a Tool for Finding Linked Genes and Detecting that Difference from Allelic Association between Loci, Proc. Nat. Acad. Sci.* USA, **85**:9119-9123,(1998)

12Tang, H.; Coram, M.; Wang, P.; Xiaofeng, Z. and Risch, N.; *Reconstructing Genetic Ancestry Blocks in Admixed Individuals Am. J. Hum. Genet.*. **79**, p.1-12,(2006)

13 Cardon, L. R. and Palmer, L. J., *Population stratification and spurious allelic association, Lancet,* **361**:598-604, (2003)

14 Jorde, L.B., *Linkage Disequilibrium and the Search for Complex Disease Genes, Genome Res.* **10**: 1435-1444, (2000)

15 W.G., Robertson A.R., *Linkage disequilibrium in finite populations, Theor. Appl. Genet.* **38**:226-231,(1968).

16. C-K, Buldyrev SV, Havlin S, Simons M, Stanley HE, Goldberger AL., *Mosaic organization of DNA nucleotides, Phys Rev E*;49:1685-1689, (1994).

17. C-K, Havlin S, Stanley HE, Goldberger AL. *Quantification of scaling exponents and crossover phenomena in nonstationary heartbeat time series. Chaos* 5:82-87 (1995)

18. W., Holste, D., *Universal 1/f noise, crosovers of scaling exponents, and chromosome-specific patterns of guanine-cytosine content in DNA sequences of the human genome, Physical Review E*, 71, 0419410-19, (2005)

19. nagel,M., Fürst, R., Rhode, K., *Entropy as a Measure for Linkage Disequilibrium over Multilocus Haplotype Blocks, Hum Hered* 54, 186-198, (2002).

20. AC, Hseu SS, Yien HW, Goldberger AL, Peng CK, *Linguistic analysis of the human heartbeat using frequency and rank order statistics, Phys Rev Lett* 90: 108103, (2003).

21. Zipf, *Human Behavior and the Principle of Least Effort, Addison-Wesley Press Inc.*, Cambridge, (1949).

22. Mandelbrot, *An informational theory of the statistical structure of languages*, in *Communication Theory*, ed. W. Jackson, Betterworth, (1953) 486.

23. Stanley,H. E., Buldyrev,S.V., Goldberger, A.L., Havlin, S.,Peng, C.-K. and Simons, M., *Scaling Features of Noncoding DNA, Physica A* 273, 1-18, (1999).

55

24. Cantú-Bolán, B., Hernández-Lemus, E., *Statistical properties and linguistic coherence in noncoding DNA sequences*, Rev. Méx. Fis. E 51 (2) 118-125, (2005).

25. Dagdug, L., Alvarez-Ramirez, J., Lopez, C., Moreno, R., and Hernández-Lemus, E., (2007); *Correlations in a Mozart's Music Score (K-73x) with Palindromic and Upside-Down Structure, Physica A* 383, 570-584, (2007).

26. Smith, T.G., W.B. Marks, G.D. Lange, W.H. Sheriff and E.A. Neale, *A fractal analysis of cell images. J.Neurosci. Meth.* 27: 173-180 (1989)

27. Fitter, A.H. and T.R. Strickland, *Fractal characterization of root system architecture, Funct. Ecol.* 6: 632-635 (1992)

28. Vlcek, J. and E. Cheung, *Fractal analysis of leaf shapes, Can. J. For. Res.* 16: 124-127, (1986)

29. Sugihara, G., *Nonlinear forecasting for the classification of natural time series, Phil. Trans. R. Soc. Lond. A*: 477-495, (1994), see also Sugihara, G. and R.M. May, *Applications of fractals in ecology, Trends Ecol. Evol.* 5: 79-86 (1990) and Sugihara, G., B. Grenfell and R.M.May, *Distinguishing error from chaos in ecological time series, Phil.Trans. R. Soc. London B* 330: 235-251 (1990).

30. Nelson, T.R., B.J. West and A.L.Goldberger, *The fractal lung: universal and species-related scaling patterns, Experimentia* 46: 251-254, (1990)

31. Barbará, D., *Chaotic Mining: Knowledge Discovery Using the Fractal Dimension, Proceedings of the 1999 SIGMOD DMKD Workshop*, Philadelphia, PA, June (1999)

32. Ruelle, D., *Chaotic Evolution and Strange Attractors*, Accademia Nationale dei Lincei, Cambridge University Press, Cambridge (1989).

33. P. Grassberger and I. Procaccia, *Characterization of Strange Attractors, Phys. Rev. Lett.* 50, 346 - 349 (1983), see also P. Grassberger and I. Procaccia, *Estimation of the Kolmogorov entropy from a chaotic signal, Phys. Rev. A* 28, 2591 (1983)

34. Scheet, P. and Stephens, M., *A fast and flexible statistical model for large-scale population genotype data: applications to inferring missing genotypes and haplotypic phase, Am J Hum Genet*, 78, 629-644, (2006)

35. Barrett, J. C., Fry, B. Müller, J. and Daly, M. J., *Haploview: Analysis and visualization of LD and haplotype maps, Bioinformatics*, **21**: 2, 263-265,(2005).

36. www.physionet.org/physiotools/

37. Sarraile, J.J., Myers, L.S., *FD3: A Program for Measuring Fractal Dimension, Educational and Psychological Measurement*, Vol. 54, No. 1, 94-97 (1994)

38.Liebovitch, L.S. and Toth, T.; *A fast algorithm to determine fractal dimensions by box counting, Physics Letters A*, 141, 386-390 (1989)
39. Twells, R.C.J., Mein,C.A., Phillips, M. S., Hess, J.F., Veijola, R., Gilbey, M., Bright, M., Metzker, M., Lie,B.A., Kingsnorth, A., Gregory, E., Nakagawa, Y., Snook, H., Wang,W.Y.S., Masters, J., Johnson, G., Eaves, I., Howson,J.M.M., Clayton,D., Cordell, H.J., Nutland,S., Rance, H., Carr, P., Todd, J.A., *Haplotype Structure, LD Blocks, and Uneven Recombination Within the LRP5 Gene, Genome Res.* Vol 13, Issue 5, 845-855, May (2003)

Theoretical Basis For The Marmur-Doty Relation

Leonardo Dagdug

Departamento de Fisica, Universidad Autonoma Metropolitana-Iztapalapa, 09340, Mexico DF, Mexico.
dll@xanum.uam.mx

Abstract. The well known linear relation between the base composition of DNA, expressed in terms of percentage of guanine plus cytosine bases and the denaturation temperature is predicted. To this end we propose a theoretical model to describe the denaturation processes of DNA as a Markov processes, taking into account the probability of finding each nearest stacked neighbor with or without their hydrogen bonds. Our main result is the theoretical prediction of the slop in the Marmur-Doty equation. This model extends the stochastic matrix method used to describe the glass transition in strong glasses to the study of configurational changes in the denaturation process.

Keywords: DNA, melting temperature.
PACS: 87.10+e

INTRODUCTION

Recently the Human Genome Project has provided us a catalog of tens of thousands of genes, now an important problem is to develop appropriate tools to understand and use this information. In particular, biological processes such as DNA replication, transcription, translation, mutation and repair are of huge importance. The statistical thermodynamics of melting DNA has been studied for a number of years. However, despite the fact that more than a half century has passed since the discovery of the structure of the DNA double helix, major questions still remain regarding its thermodynamic behavior and stability. Thus, the parameters that determine the cooperativity of melting have been difficult to measure and also it has been difficult to test some basic theoretical assumptions such as the role of heat capacity changes, volume changes and compressibility, that accompany nucleic acid conformational transitions [1].

One of the ways to learn about the structure of macromolecules in solution is to observe structural changes. The ordered form of a nucleic acid is only marginally stable against temperature increase, so that most samples show a drastic alteration in structure between the convenient limits of 0°-100 ° C. Many physical properties are changed in the process, and the nature of these changes and characteristics of the transformation provide fertile ground for physical studies [2]. The most common method of following the denaturation of DNA is the profile of ultraviolet absorbance

CP978, *Biological Physics, 3rd Mexican Meeting on Mathematical and Experimental Physics*
edited by L. Dagdug and L. García-Colín Scherer
© 2008 American Institute of Physics 978-0-7354-0497-7/08/$23.00

against temperature, called the melting curve. An important quantity is the characteristic transition temperature T_m. T_m is defined as the temperature at which half of the strands are in the double-helical state and the other half are in the "random-coil" state. A DNA melting curve is generally a two-dimensional plot displaying some properties of a DNA solution against an external variable producing DNA unwinding. The most common external variable is the temperature, but the process can also be observed at extremes of pH, decreases in the dielectric constant of the aqueous medium, and exposure to amides, urea and similar solvents. The DNA property of optical absorbance can be monitored at approximately 260 nm, while increasing the temperature and normalizing the absorbance change in an appropriate way. Thus, a plot of the fraction of broken base-pairs (bps) versus temperature is obtained. In the double helix, disruption of the ordered state, with its stacked base pairs, leads to less frequent contact between the bases and an increased absorbance [3]. Since the pioneering work of Zimm [4], and the appearance of the nearest-neighbor (NN) model, several theoretical and experimental papers on DNA thermodynamics have appeared. These works provide the complete thermodynamics library of all 10 Watson-Crick DNA nearest-neighbor interactions. Good descriptions of a sampling of experimental techniques used for this purpose and the principal thermodynamics libraries, are described in [5]. Even thought the thermodynamics sets given in these articles disagree a number of issues, they show how these thermodynamics data can be used to calculate the stability of the structure from knowledge of its base sequence. The particular differences are between DNA polymer and oligonucleotides and in the salt dependence of nucleic acid denaturation. As result of these works, it is well established that[1] [6]

$$\uparrow \begin{matrix} Py{\cdot}Pu \\ Pu{\cdot}Py \end{matrix} \downarrow$$

is more stable than

$$\uparrow \begin{matrix} Pu{\cdot}Py \\ Py{\cdot}Pu \end{matrix} \downarrow$$

Also, theoretical efforts have been made to calculate stacking energies *ab initio*; for a review consult reference [7] The first theoretical attempt to model this transition was a one-dimensional Ising-like model in which the two states of spin correspond to an open or closed state of the base pair, with a favorable coupling between neighbor pairs that are in the same state [4]. It reproduced a crossover between the two different regimes but no thermo-dynamical transition. The relative tendencies of the system to occupy one of the two states were introduced explicitly in terms of free enthalpies an

---

[1] Py stands for pyrimidine and Pu for purine. Arrows designate the direction of the sugar-phosphate chain, from the C'_3 atom of a deoxyribose unit to the C'_5 atom of the next deoxyribose adjacent to and on either side of the phosphodiester linkage. Sometimes nearest-neighbor base pairs are represented with a slash separating strands in an antiparallel orientation (eg.,

AC / TG means $5' - AC - 3'$ Watson-Crick bases paired with $3' - TG - 5'$ or $\uparrow \dfrac{A{\cdot}T}{C{\cdot}G} \downarrow$ in the notation

used throughout this paper).

their temperature dependence. Although the difficulties in choice of such enthalpies the method has proved useful in describing some aspects of DNA denaturation [8]. Understanding of this remarkable one-dimension cooperative phenomenon in terms of Hamiltonian model with independent parameters remained an outstanding problem. Recent research has emphasized the role of the large amplitude fluctuations that precede the transition and the intrinsically nonlinear mechanisms, which are needed to describe such fluctuations [8]. This description was introduced because experimentally, a purified DNA sample containing a unique sequence and length is found to exhibit distinct multi step melting [3].

In 1962 Marmur and Doty discovered a linear relation between the base composition of DNA [9], expressed in terms of percentage of guanine plus cytosine bases and the denaturation temperature, T_m, given by $T_m = 69.3 + 0.41(G - C)$. In this relation the temperature is given in Centigrade and $G - C$ refers to the mole percentage of guanine plus cytosine and goes from 30 to 80%. It is also well known that while the abscissa depends on the solvent containing, the slop is a universal constant. In the present paper we propose a theoretical model to describe the denaturation processes of DNA as a Markov processes, taking into account the probability of finding each nearest stacked neighbor with or without the hydrogen bonds, our main result is the theoretical prediction of the slop in the Marmur-Doty equation. This model extends the stochastic matrix method (SMM) [10], used to describe the glass transition in strong glasses, to the study of configurational changes in the denaturation process and to the prediction of T_m of DNA.

MODEL AND DEFINITIONS

In our theoretical model the statistical treatment is based on a first-order Markov process and the fact that the thermodynamic values for the ten possible dimers are known. As a consequence the external conditions as pH, salt concentration, solution and the role of the large amplitude fluctuations and the intrinsically nonlinear mechanisms are included on them.

In the stochastic matrix method, the process of observing the configuration of bps in DNA can be described by a matrix (M) acting on an initial vector v_0 (which characterizes the initial condition of the bps), if the matrix components are the probability of having a bp (bp) neighboring another one in a specific configuration. The probability of having some configuration of bps is modeled by n successive applications of the matrix M on an initial vector v_0. After n applications, the final configuration of the system can be written as a linear combination of the eigenvectors associated with M, i. e., $v_n = \sum_{m=1} a_m \lambda_M^n e_m$ where e_m is the eigenvector

M with eigenvalue λ_M^n, and a_m is the projection of v_0 onto e_m.

A matrix with all the columns normalized to one, as M, has the property that at least one eigenvalue is one, while the real part of all of the rest is less than one. This result allows us to assert that only the eigenvectors with eigenvalues equal to one survive after successive applications of M onto v_0. If we assume that M has one eigenvector e_1 with eigenvalue equal to one, then in the limit of large n, v_∞ converges to e_1, with $a_1 = 1$, due to a conservation of probability. Therefore, this means that the configuration attains a steady statistical regime represented by e_1. The explicit form of this eigenvector is obtained by solving the system of equations:

$$(M-1)e_1 = 0 \qquad (1)$$

which enables us to calculate the probability of any configuration in the system.

To construct the stochastic matrix describing the melting behavior of DNA, we first need to define the units. These units must be given by four combinations: A and T, T and A, G and C, and C and G. The bps can be bonded or unbonded. This can be represented as $\uparrow$A•T$\downarrow$, $\uparrow$T•A$\downarrow$, $\uparrow$G•C$\downarrow$, $\uparrow$C•G$\downarrow$, $\uparrow$AT$\downarrow$, $\uparrow$TA$\downarrow$, $\uparrow$GC$\downarrow$, $\uparrow$CG$\downarrow$, where the dot represents the existence of the hydrogen bonds between the bps, and the absence of the dot represents the unbonded bps. These eight units give 16 different combinations of base-pair stacking, where each site of the matrix represents the probability of finding a specific configuration of each duplex. The 16 different combinations can be displayed as a 4×4 matrix, namely:

$$
\begin{pmatrix}
c^2 \uparrow \begin{smallmatrix} \text{G•C} \\ \text{G•C} \end{smallmatrix} \downarrow & c(1-c) \uparrow \begin{smallmatrix} \text{A•T} \\ \text{G•C} \end{smallmatrix} \downarrow & c^2 \uparrow \begin{smallmatrix} \text{G C} \\ \text{G•C} \end{smallmatrix} \downarrow & c(1-c) \uparrow \begin{smallmatrix} \text{A T} \\ \text{G•C} \end{smallmatrix} \downarrow \\
c(1-c) \uparrow \begin{smallmatrix} \text{G•C} \\ \text{A•T} \end{smallmatrix} \downarrow & (1-c)^2 \uparrow \begin{smallmatrix} \text{A•T} \\ \text{A•T} \end{smallmatrix} \downarrow & c(1-c) \uparrow \begin{smallmatrix} \text{G C} \\ \text{A•T} \end{smallmatrix} \downarrow & (1-c)^2 \uparrow \begin{smallmatrix} \text{A T} \\ \text{A•T} \end{smallmatrix} \downarrow \\
c^2 \uparrow \begin{smallmatrix} \text{G•C} \\ \text{G C} \end{smallmatrix} \downarrow & c(1-c) \uparrow \begin{smallmatrix} \text{A•T} \\ \text{G C} \end{smallmatrix} \downarrow & c^2 \uparrow \begin{smallmatrix} \text{G C} \\ \text{G C} \end{smallmatrix} \downarrow & c(1-c) \uparrow \begin{smallmatrix} \text{A T} \\ \text{G C} \end{smallmatrix} \downarrow \\
c(1-c) \uparrow \begin{smallmatrix} \text{G•C} \\ \text{A T} \end{smallmatrix} \downarrow & (1-c)^2 \uparrow \begin{smallmatrix} \text{A•T} \\ \text{A T} \end{smallmatrix} \downarrow & c(1-c) \uparrow \begin{smallmatrix} \text{G C} \\ \text{A T} \end{smallmatrix} \downarrow & (1-c)^2 \uparrow \begin{smallmatrix} \text{A T} \\ \text{A T} \end{smallmatrix} \downarrow
\end{pmatrix} \qquad (2)
$$

where

$$\uparrow \begin{smallmatrix} \text{G•C} \\ \text{G•C} \end{smallmatrix} \downarrow$$

represents the probability of having a bonding $\uparrow$G•C$\downarrow$ neighboring a bonding $\uparrow$G•C$\downarrow$, and

$$\uparrow \begin{smallmatrix} \text{G•C} \\ \text{G C} \end{smallmatrix} \downarrow$$

represents the probability of having a bonding $\uparrow$G•C$\downarrow$ neighboring an unbonded $\uparrow$G C$\downarrow$, etc. c is de fraction of G+C in the DNA chain, and (1-c) is the fraction of

A + T. These stacking processes are in three dimensions, and this information must be included in the stacking energy, as should the information concerning the properties of the solvent in which the transition is carried out. Based on the NN model, each configuration is proportional to two factors: the concentration of the bps and its Boltzmann factor. The first one depends on the sequence, and the second one, on the possible configurations. This last factor involves the Gibbs free energy of each configuration, namely, $\exp(\Delta G_{MN} / k_B T)$ (where M and N stand for A, T, G, C, and MN represents the stacked bps in a single strand in the direction $5' - 3'$). If each configuration is proportional to its stability constant, it means that if the probability of some bp is found bond or unbonded at a given temperature, this bp is unable to change its configuration at this temperature.

The eigenvector with eigenvalue equal to one of matrix (2) is a vector with eight components that gives us the probability of finding at a fixed temperature, the following configurations in the system: $\uparrow$A·T$\downarrow$, $\uparrow$T·A$\downarrow$, $\uparrow$G·C$\downarrow$, $\uparrow$C·G$\downarrow$, $\uparrow$A T$\downarrow$, $\uparrow$T A$\downarrow$, $\uparrow$G C$\downarrow$, and $\uparrow$C G$\downarrow$. The sum $\uparrow$G C$\downarrow$ + $\uparrow$A T$\downarrow$ give us the probability of denatured bps in the system. Setting this sum to $1/2$, T_m is obtained for DNA.

In the next section we will use matrix (2) to obtain the behavior of the melting transition and.

RESULTS

To calculate the T_m we proceed as follows. Obtaining the eigenvector with eigenvalue equal to one, we can find the probability of natured and denatured bps. Using the definition of T_m, we set the probability of denatured bps to 1/2, and finally we solve the equation for T_m to find the melting temperature.

Inserting the energetic contributions in matrix (2) we find that,

$$
\begin{pmatrix}
c^2\xi & c(1-c)\varepsilon & c^2\xi\mu_1 & c(1-c)\varepsilon\mu_2 \\
c(1-c)\eta & (1-c)^2\lambda & c(1-c)\eta\mu_1 & (1-c)^2\lambda\mu_2 \\
c^2\xi\mu_1 & c(1-c)\varepsilon\mu_1 & c^2 & c(1-c) \\
c(1-c)\eta\mu_2 & (1-c)^2\lambda\mu_2 & c(1-c) & (1-c)^2
\end{pmatrix}
\tag{3}
$$

where $\xi \equiv \exp(\Delta G_{GG} / k_B T)$, $\eta \equiv \exp(\Delta G_{GA} / k_B T)$, $\varepsilon \equiv \exp(\Delta G_{AG} / k_B T)$ and $\lambda \equiv \exp(\Delta G_{AA} / k_B T)$ are the free energy for the each dimmer. $\mu_1 \equiv \exp(\Delta G_{GCh} / k_B T)$ and $\mu_2 \equiv \exp(\Delta G_{ATh} / k_B T)$ involves the free energy for the hydrogen bond between G and C and A and T respectively. If the hydrogen bond of a bp is broken in a duplex, we have $\Delta G_{MNh} - \Delta G_{MNh}$, and the Boltzmann factor for this configuration is given by $\xi\mu_1$, etc. If two hydrogen bonds of the bps involved are broken, its Gibss free energy is zero. Some approximations can be done to simplify the algebra to manipulate

61

matrix (3). First of all, the experimentally is well known that $\Delta G_{\text{ATh}} \approx 4 Kcal/mol$ and $\Delta G_{\text{GCh}} \approx 6 Kcal/mol$, and because those are divided by kT in each Boltzamn factor and the temperature is around room temperature, both can be taken as his average value, $\Delta G_{\text{GCh}} \approx \Delta G_{\text{ATh}} \approx 5 Kcal/mol$. So in matrix (3) we can take $\mu_1 = \mu_2 = \mu$. Secondly, also is well know that $\xi \approx \varepsilon$ and $\eta \approx \lambda$. Under this approximation matrix (3) can be written as

$$
\begin{pmatrix}
c^2\xi & c(1-c)\xi & c^2\xi\mu & c(1-c)\xi\mu \\
c(1-c)\eta & (1-c)^2\eta & c(1-c)\eta\mu & (1-c)^2\eta\mu \\
c^2\xi\mu & c(1-c)\xi\mu & c^2 & c(1-c) \\
c(1-c)\eta\mu & (1-c)^2\eta\mu & c(1-c) & (1-c)^2
\end{pmatrix}
\tag{4}
$$

After normalizing each column of matrix ((4)) we have:

$$
\begin{pmatrix}
c^2\xi\alpha & c(1-c)\xi\alpha & c^2\xi\mu\beta & c(1-c)\xi\mu\beta \\
c(1-c)\eta\alpha & (1-c)^2\eta\alpha & c(1-c)\eta\mu\beta & (1-c)^2\eta\mu\beta \\
c^2\xi\mu\alpha & c(1-c)\xi\mu\alpha & c^2\beta & c(1-c)\beta \\
c(1-c)\eta\mu\alpha & (1-c)^2\eta\mu\alpha & c(1-c)\beta & (1-c)^2\beta
\end{pmatrix}
\tag{5}
$$

where $\alpha^{-1} = (1+\mu)(c\varepsilon + (1-c)\lambda)$ and $\beta^{-1} = \mu(\lambda(1-c) - c\varepsilon) + 1$.

The explicit form of the eigenvector with eigenvalue one is obtained by solving equation (1) with M given by matrix (5), and a vector of four components is found. The first and second components of this vector give us the probability of finding the closed duplex. The sum of the last two terms gives us the probability of finding the totally open bps, namely,

$$
\theta = \frac{1 + c(\varepsilon - \lambda)\mu + \lambda\mu}{1 + \lambda + 2\lambda\mu + c(\varepsilon - \lambda)(1 + 2\mu)}
\tag{6}
$$

Equation (6) allows us to calculate the fraction of open bps. depending on temperature, and is our main result.

To calculate T_m, by definition, it is necessary to impose $\theta = 1/2$. Equation (6) is equal to $1/2$ when,

$$
\lambda + c(\varepsilon - \lambda) - 1 = 0
\tag{7}
$$

Equation (7) depends on the hydrogen bond parameters as well as the stacking ones. In this case, the role of the stacking energies is as fundamental as that of the hydrogen bonding energies, and the competition between $\Delta S's$ and $\Delta H's$ governs the melting behavior. The theoretical values for T_m are obtained using equation (7) and the experimental nearest-neighbor thermodynamics set of data are from Breslauer (Table 2 Ref. [11]), for an excellent description and discussion about the thermodynamics libraries see reference [5]). This set of data is used because it has been shown to be the best one [12]. In Fig. 2 is plotted T_m vs. c from equation (7). The slop obtained by the best linear fit by the theoretical data calculated from equation (6), when the denatured temperature is plotted vs. guanine plus cytosine is 0.39, less than 5% respect to the experimental value reported by Marmur and Doty (0.41).

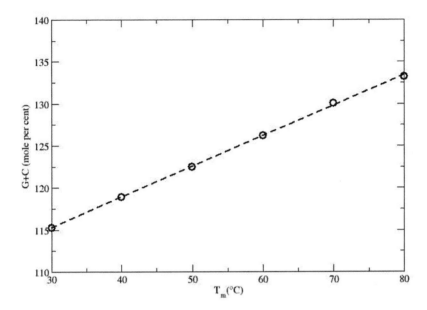

FIGURE 1. Dependence of the denaturation temperature, T_m, on guanine plus cytosine content of DNA calculated by equation (6). The dashed line is the best linear fit of the theoretical data.

Studding the limiting concentration case with our analytical solutions is simple. When $c = 1$ and $c = 0$, equation (6) is reduced to

$$\theta_{c=0} = \frac{1 + \lambda\mu}{1 + \lambda + 2\lambda\mu} \tag{8}$$

and

$$\theta_{c=1} = \frac{1 + \varepsilon\mu}{1 + \varepsilon + 2\varepsilon\mu}. \tag{9}$$

On the other hand, equation (7) goes to $\lambda = 1$ and $\varepsilon = 1$ when $c = 1$ and $c = 0$ respectively. This means that $T_m = \Delta H_{NM}/\Delta S_{NM}$, and the melting temperature only depends on one thermodynamic staking bp parameter, as should be.

CONCLUSIONS

To summarize, we used the stochastic matrix method to study the DNA melting temperature dependence on the guanine plus cytosine fractional composition and the DNA melting. The elements of the stochastic matrix are the probabilities that any base-pair has a bonded or unbonded neighbor. If ones assume that the stochastic matrix has an eigenvector with eigenvalue equal to one, the possible configurations of the system are fixed by this eigenvector. In fact, this vector is the probability of finding any base bonded or unbonded. Once we have this probability, we are able to obtain an analytic expression for the fraction of broken base pairs as a function of temperature. Also, setting to 1/2 the expression for this probability (equation (6)), gives us an analytical expression to calculate the melting temperature. This theoretical method is free of adjustable parameters, and to carry out the comparison with experimental data, only the DNA nearest-neighbor thermodynamics set, as well as the hydrogen bond parameter is needed. The slop predicted by our theoretical expression for temperature dependence of the denatured base-pairs are on the 5% error.

REFERENCES

1. R. A. Dimitrov and M. Zuker, *Biophysical Journal* **87**, 215-226 (2004).
2. V. A. Bloomfield, "Physical chemistry of nucleic acids", Harper and Row, New York, 1974.
3. M. Wartell and A. S. Benight, *Phys. Rep.* **126** , 67-107 (1985).
4. D. M. Crothers and B. H. Zimm, *J. Mol. Biol.* **9**,1-9 (1964).
5. Jr. SantaLucia, *Proc. Natl. Acad. Sci. USA.* **95**, 1460-1465 (1998).
6. W. Saenger, in Principles of Nucleic Acids Structure, (Spring-Verlag, New York, 1983).
7. P. Hobza and J. Sponer, *Chem. Rev.* **99**, 3247-3276 (1999).
8. D. Poland and H. A. Sheraga, *J. Chem. Phys.* **45**, 1456-1464 (1966).
9. J. Marmur and P. Doty, *J. Mol. Biol.* **5**, 109-118 (1962).
10. R. Kerner, *Phys. B.* **215**, 267-272 (1995).
11. K. J. Breslauer, *Proc. Natl Acad. Sci. USA.* **83**, 3746-3750 (1986).
12. P. Miramontes and G. cocho, *Phys. A.* **321**, 577-586 (2003).

Using non-equilibrium simulations to estimate equilibrium binding affinities

F. Marty Ytreberg

Department of Physics, University of Idaho, Moscow, Idaho 83844

Abstract. We demonstrate that non-equilibrium unbinding simulations can be used to accurately estimate equilibrium binding affinities. Utilizing the FKBP protein bound to two different ligands we estimate the absolute binding affinities within less than 1.0 kcal/mol of the experimental values. The methodology is straight-forward, requiring no modification to many modern molecular simulation packages. The approach makes use of a physical pathway, eliminating the need for complicated alchemical decoupling schemes. These results suggest that non-equilibrium simulation could provide a viable means to accurately estimate protein-ligand binding affinities.

Keywords: Free energy, Jarzynski equality
PACS: 87.15.kp, 87.15.ap

INTRODUCTION

The accurate estimation of binding affinities for protein-ligand systems (ΔG) remains one of the most challenging tasks in computational biophysics and biochemistry. Due to the high computational cost of such free energy computation, it is of interest to understand the advantages and limitations of various ΔG methods.

Many previous studies (e.g., Refs. [1–21]) have calculated protein-ligand binding affinities using equilibrium free energy methods such as thermodynamic integration [22], free energy perturbation [23, 24], and weighted histogram analysis [25]. Due to the recently introduced Jarzynski approach [26] it is also possible to estimate ΔG from non-equilibrium simulations. However, the estimation of ΔG for protein-ligand binding using non-equilibrium approaches remains largely untested. One such study was performed by the McCammon group and is detailed in Ref. [27]. However, no experimental data were available, and thus the accuracy of the calculations could not be discussed. Another study by the Grubmüller group found that non-equilibrium simulations resulted in a large overestimation of ΔG as compared to experiment [28]. Their conclusion was that the ΔG estimate was not fully converged.

In this report we demonstrate the ability to compute accurate (as compared to experimental data) ΔG estimates following a non-equilibrium methodology. The approach relies on performing multiple non-equilibrium unbinding simulations using a physical pathway, i.e., pulling the ligand out of the binding pocket, and then uses the Jarzynski relation [26] to estimate ΔG. The system is an FKBP protein complexed with 4-hydroxy-2-butanone (BUQ) and dimethyl sulfoxide (DMSO). The motivation for using this system is that comparison to experiment is possible [29] and many previous computational studies have been performed [11, 12, 17, 19, 20]. Our results are encouraging—ΔG estimates are within less than 1.0 kcal/mol of the experimental value for each ligand.

CP978, *Biological Physics, 3rd Mexican Meeting on Mathematical and Experimental Physics*
edited by L. Dagdug and L. García-Colín Scherer
© 2008 American Institute of Physics 978-0-7354-0497-7/08/$23.00

The importance of pursuing non-equilibrium methods such as used in this report is three-fold: (i) The approach is trivially parallelizeable since each non-equilibrium unbinding simulation is performed independently. (ii) The method is simple to implement in many existing simulation packages such as GROMACS [30], used here; no modification to the code is necessary. (iii) Since a physical pathway is utilized, there is no need to use complicated alchemical decoupling schemes as is often the case with ΔG computation.

THEORY

In general, the absolute binding affinity is defined as the free energy difference between the unbound (apo) and bound (holo) states of the protein-ligand system. We define the apo state as when the protein and ligand are not interacting due to a large separation between them. The holo state is defined by the ligand in the binding pocket of the protein. If the system is described by a potential energy function $U(\vec{x})$, where $\vec{x}$ is the full set of configurational coordinates, then the absolute binding affinity $\Delta G_{\text{bind}} = \Delta G_{\text{apo}\rightarrow\text{holo}}$ can be written in terms of the partition functions for each state $e^{-\beta \Delta G_{\text{bind}}} = Z_{\text{holo}}/Z_{\text{apo}}$, where $Z_S = \int_{\vec{x} \in S} d\vec{x}\, e^{-\beta U(\vec{x})}$ is the configurational partition function for state S, and $\beta = 1/kT$ is the inverse system temperature. Because the apo and holo states are configurationally very different from one another, a pathway connecting them is created. For the current study this pathway is a physically viable route that moves the ligand from the binding pocket to a large distance away from the protein.

To carry out the non-equilibrium unbinding simulations we pulled the ligand out of the binding pocket using a method analogous to atomic force microscopy (compare to Refs. [27, 28, 31–33]). Thus, a spring was attached to the center of mass of the ligand and then moved at a constant speed and direction out of the binding pocket, pulling the ligand away from the protein. The protein was also restrained by attaching a stationary spring to its center of mass.

The approach used here makes use of the well-known Jarzynski equality [26, 34, 35]. For the unbinding simulations considered here, the work values W are used to generate the potential of mean force (PMF) as a function of the protein-ligand separation r using the stiff spring approximation [32, 33]

$$e^{-\beta \Phi(r)} \approx e^{-\beta \Delta G(r)} = \left\langle e^{-\beta W(r)} \right\rangle_{\text{holo}}, \tag{1}$$

where $\Phi(r)$ is the PMF, and the notation $\langle ... \rangle_{\text{holo}}$ is a reminder that the equality holds only when all realizations of the work W have been obtained, and that each W must be generated by starting the system from a structure in the equilibrium ensemble for the holo state.

For the results given in this report the ligand is pulled out of the binding pocket, and the reverse process of pulling the ligand into the pocket is not considered. Although the use of bi-directional simulation has been shown to be an effective approach to accurate ΔG estimation [36–42], the results for pulling the ligand into the pocket would be unreliable since the ligand would have to find the most important binding pose(s) during

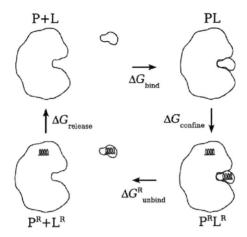

P+L PL

ΔG_{bind}

$\Delta G_{\mathrm{release}}$ $\Delta G_{\mathrm{confine}}$

$\Delta G^{\mathrm{R}}_{\mathrm{unbind}}$

$P^{R}+L^{R}$ $P^{R}L^{R}$

FIGURE 1. Thermodynamic cycle used to compute the binding affinity ΔG_{bind} using non-equilibrium unbinding simulations. The springs in the diagram and the "R" superscript are to remind the reader that the protein and ligand center of mass are restrained during the unbinding process.

the course of the simulation. In addition, the protein could rotate during the simulation making it difficult for the ligand to enter the binding pocket.

We note two aspects of the relationship embodied in Eq. (1): (i) The equality holds only in the case of obtaining all possible work values W. Thus, it is important to calculate uncertainty estimates for ΔG, and if possible, to compare results to experimental data. (ii) The relation is independent of the speed at which the system is forced, i.e., the unbinding speed. In practice, however, it has been found that the speed chosen can dramatically affect the convergence behavior of the ΔG estimate [42–44].

Absolute binding affinity calculation

Figure 1 shows the thermodynamic cycle used for this study. The desired binding affinity that can be compared to experiment is given by ΔG_{bind}. The unbinding free energy obtained by use of the Jarzynski relation and the PMF is given by $\Delta G^{\mathrm{R}}_{\mathrm{unbind}}$ where the superscript in the notation and the spring in the figure are to remind the reader that both the ligand and protein are restrained during the unbinding event. Since free energy is a state function, the figure gives

$$\Delta G_{\mathrm{bind}} = -\Delta G_{\mathrm{release}} - \Delta G^{\mathrm{R}}_{\mathrm{unbind}} - \Delta G_{\mathrm{confine}}. \qquad (2)$$

The quantities $\Delta G_{\mathrm{confine}}$ and $\Delta G_{\mathrm{release}}$ in Eq. (2) can be interpreted as, respectively, the free energy change associated with confining the protein and ligand with restraints in the holo state, and the free energy of restraint release in the apo state. Each of these terms can be further divided into two contributions: one for the protein $\Delta G^{\mathrm{P}}_{\mathrm{release}}$ and another for the ligand $\Delta G^{\mathrm{L}}_{\mathrm{release}}$. So, $\Delta G_{\mathrm{release}} = \Delta G^{\mathrm{P}}_{\mathrm{release}} + \Delta G^{\mathrm{L}}_{\mathrm{release}}$ with a similar form for $\Delta G_{\mathrm{confine}}$.

First, we note that $\Delta G_{\text{release}}^{\text{P}} \approx -\Delta G_{\text{confine}}^{\text{P}}$ since the surrounding environment for the protein is the same for both apo and holo conformations. Secondly, the free energy to confine the ligand using a harmonic restraint applied to the center of mass can be obtained analytically [2, 15, 17]

$$\Delta G_{\text{confine}}^{\text{L}} = -kT \ln \left(\frac{V_{\text{rest}}^{L}}{V_{\text{free}}^{L}} \right), \quad V_{\text{rest}}^{L} = \left(\frac{2\pi kT}{k_s} \right)^{3/2}, \tag{3}$$

where V_{rest}^{L} and V_{free}^{L} are the molecular volumes for the restrained and unrestrained ligand respectively, and k_s is the spring constant used for the restraint. For the holo state, since the ligand is confined to the binding pocket, the restraint has little affect on the effective volume available to the ligand, i.e., $V_{\text{rest}}^{L}(\text{holo}) \approx V_{\text{free}}^{L}(\text{holo})$, and thus $\Delta G_{\text{confine}}^{\text{L}} \approx 0$. Also, we note that in the apo state the free volume is given by the standard concentration, i.e., $V_{\text{free}}^{L}(\text{apo}) = V_0 \approx 1.661 \text{ nm}^3$; and thus $\Delta G_{\text{release}}^{\text{L}} = +kT \ln \left[(2\pi kT/k_s)^{3/2}/V_0 \right]$.

It has been shown that the free energy difference $\Delta G_{\text{unbind}}^{\text{R}}$ can be obtained by integrating the PMF (See discussion in Refs. [3, 15, 20])

$$e^{-\beta \Delta G_{\text{unbind}}^{\text{R}}} = \frac{4\pi r_{\text{ref}}^2}{V_0} \int dr \, e^{-\beta (\Phi(r) - \Phi(r_{\text{ref}}))}, \tag{4}$$

where r_{ref} is a reference protein-ligand separation, chosen to be large enough that the PMF has plateaued indicating that the protein and ligand are not interacting. Thus, the absolute binding affinity can be written as

$$\Delta G_{\text{bind}} \approx -\Delta G_{\text{unbind}}^{\text{R}} - \Delta G_{\text{release}}^{\text{L}}$$

$$\approx -kT \ln \left[\frac{4\pi r_{\text{ref}}^2}{V_0} \int dr \, e^{-\beta (\Phi(r) - \Phi(r_{\text{ref}}))} \right] - kT \ln \left[\frac{1}{V_0} \left(\frac{2\pi kT}{k_s} \right)^{3/2} \right]. \tag{5}$$

All binding free energy data given below uses Eq. (5).

METHODS

Computational details

The initial coordinates for the FKBP-ligand complexes were obtained from the Protein Data Bank [45]: 1D7H for FKBP-DMSO, and 1D7J for FKBP-BUQ. The topologies for DMSO and BUQ were then generated by the PRODRG server [46], with partial charges slightly modified by the author.

The GROMACS simulation package version 3.3.1 [30] was used with the default GROMOS-96 43A1 forcefield [47]. Protonation states for the histidine residues were selected by the GROMACS program pdb2gmx: HIS-25 was protonated at $N\delta 1$, and HIS-87 and HIS-94 were protonated at $N\varepsilon 2$. The protein-ligand complexes were then solvated in a cubic box of SPC water [48] of approximate initial size 6.8 nm a side. A single chloride ion was randomly placed in each water box to give a net neutral charge, and then each system was minimized using steepest decent for 500 steps. To

68

allow for equilibration of the water, each system was then simulated for 1.0 ns with the positions of all atoms in the ligand and protein harmonically restrained. The temperature was maintained at 300 K using Langevin dynamics [49] with a friction coefficient of 1.0 amu/ps. The pressure was maintained at 1.0 atm using the Berendsen algorithm [50]. We note that the Berendsen algorithm does not produce canonically distributed structures, however, it is not important for this position restrained stage since none of the simulation frames are used for producing ΔG results. The LINCS algorithm [51] was used to constrain hydrogens to their ideal lengths and heavy hydrogens were used—the hydrogen mass was increased by a factor of four and this increase was subtracted from the bonded heavy atom so that the mass of the system remained unchanged—allowing the use of a 4.0 fs timestep. Particle mesh Ewald [52] was used for electrostatics with a real-space cutoff of 1.0 nm and a Fourier spacing of 0.1 nm. Van der Waals interactions used a cutoff with a smoothing function such that the interactions smoothly decayed to zero between 0.75 nm and 0.9 nm. Dispersion corrections for the energy and pressure were utilized [53].

Finally, a 2.0 ns equilibrium simulation at constant temperature and volume was used to generate starting configurations for use in the Jarzynski method. Each FKBP-protein complex was simulated with parameters chosen identical to the position restrained simulation above (except for fixed volume). The size of the water box was chosen as the last configuration from the position restrained simulations.

Unbinding simulations

Starting structures for the unbinding simulations were chosen to be equally spaced within the 2.0 ns equilibrium simulation. So, if 20 starting structures were desired, then the spacing between snapshots was 100 ps. The pulling direction was chosen for each starting structure by visual inspection using the VMD software package [54]. We note that, in the limit of very slow pulling speeds, any reasonably chosen pulling direction will give the same results since the protein is allowed to rotate during the simulation and thus optimize the exit path.

The pulling simulations were performed using GROMACS 3.3.1 as above. All parameters were identical to the 2.0 ns equilibrium simulation. The center of mass of the protein was harmonically restrained to its initial location with a spring constant of 10,000.0 kJ/mol/nm^2. The ligand was connected to a spring moving at a constant velocity with a spring constant of 1000.0 kJ/mol/nm^2. The simulations were discontinued when the spring had traveled a total distance of 2.0 nm.

For the FKBP-DMSO system we tried four different ligand spring speeds: 10^{-3} nm/ps (2.0 ns per unbinding simulation), 5×10^{-4} nm/ps (4.0 ns per unbinding simulation), 2.5×10^{-4} nm/ps (8.0 ns per unbinding simulation), and 1.25×10^{-4} nm/ps (16.0 ns per unbinding simulation). We found that the two larger velocities produced unreliable results with large uncertainties (data not shown). Thus, for FKBP-BUQ we only attempted the two slower speeds.

The non-equilibrium unbinding simulations provided us with the positions of the ligand and the spring attached to the ligand for every time step. We then computed

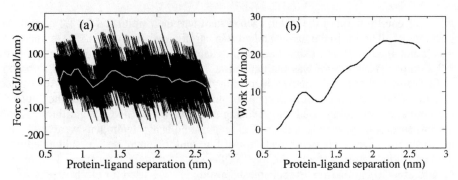

FIGURE 2. A single non-equilibrium pulling simulation performed on the FKBP-DMSO system using a pulling speed of 1.25×10^{-4} nm/ps. (a) Force on the ligand by the spring as a function of the protein-ligand separation. Both the instantaneous (black) and averaged forces (grey) are shown. (b) Work as a function of protein-ligand separation. This curve was generated by numerically integrating the averaged force in chart (a).

the force for every time step using $F(t) = k_s \left[vt - \left(\vec{\xi}(t) - \vec{\xi}_0 \right) \cdot \vec{n} \right]$, where v is the pulling speed of the spring, $\vec{n}$ is a unit vector indicating the pulling direction, $\vec{\xi}(t)$ is the position of the ligand at time t, and $\vec{\xi}_0$ is the initial position of the ligand (and spring) [27]. The center of mass separations between the protein and ligand were also computed, allowing us to generate force vs separation curves; see Fig. 2a. The forces were then averaged over intervals of 0.05 nm to give an averaged force vs separation which we then numerically integrated to obtain work vs separation; see Fig. 2b.

After the work vs separation curves were generated for each unbinding simulation desired, we used Eq. (1) to estimate the PMF as a function of the protein-ligand separation; detailed below and seen in Fig. 3b.

Uncertainty estimation

We computed the uncertainty in our ΔG_{bind} estimates using the bootstrap approach: (i) The free energy difference ΔG_{unbind}^R was estimated via Eqs. (1) and (4) using N work values chosen at random (with replacement) from a dataset containing N values. (ii) The above step was repeated until the mean and standard deviation of the free energy estimates fully converged—around 100,000 trials in our study. (iii) The uncertainty reported is the converged standard deviation of the free energy estimates.

For comparison, we also used the uncertainty analysis obtained by Zuckerman and Woolf [55], and the Bustamante group [44]. These uncertainty estimates are accurate when the variance in the estimate dominates over the bias (as is typically the case for large N).

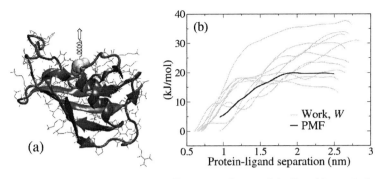

FIGURE 3. (a) The FKBP-DMSO complex. The center of mass of the ligand is attached to a spring moving at a constant velocity, and the center of mass of the protein is attached to a stationary spring. (b) The PMF (black, $\Phi(r)$ in (1) and (4)) and work values W (grey) for multiple non-equilibrium pulling simulations performed on the FKBP-DMSO system. The results are for a pulling speed of 1.25×10^{-4} nm/ps. Note that the PMF becomes approximately constant when the ligand is no longer interacting with the protein around 2.1 nm.

RESULTS AND DISCUSSION

The results of this study are very encouraging. Using the simple non-equilibrium methodology outlined above we estimated the the binding affinity for the FKBP-DMSO and FKBP-BUQ complexes within less than 1.0 kcal/mol of the experimental values.

Figure 3a shows the FKBP-DMSO complex with a sketch of the spring that is attached to the center of mass of the ligand and then moved away from the protein at constant velocity. The protein is also attached to a spring, with a larger spring constant, that is not allowed to move.

Figure 3b shows the set of work values (grey) obtained for the FKBP-DMSO system with a pulling speed of 1.25×10^{-4} nm/ps. The solid black curve shows the resulting potential of mean force (PMF) as a function of the separation between the protein and ligand, obtained via Eq. (1).

Figure 4 shows the PMF as a function of protein-ligand separation for all systems studied here. Data is shown for both DMSO and BUQ systems, with pulling speeds of 1.25×10^{-4} nm/ps and 2.5×10^{-4} nm/ps. Note that the PMF plateaus at around 2.1 nm for DMSO and around 2.3 nm for BUQ.

We estimated ΔG_{bind} to compare to experimental data using Eq. (5). Reference separations were chosen as $r_{\text{ref}} = 2.1$ nm for DMSO and $r_{\text{ref}} = 2.3$ nm for BUQ. Using $V_0 = 1.661$ nm^3, $T = 300.0$ K and $k_s = 1000.0$ kJ/mol/nm^2 the contribution from the second term in Eq. (5) is 16.81 kJ/mol for both DMSO and BUQ systems.

The binding affinity results obtained via Eq. (5) are shown in Tab. 1. Uncertainty estimates are obtained using both a bootstrap method and the approach described in Refs. [44, 55]. The computational estimates are in excellent agreement with experimental data; all results are within less than 3.0 kJ/mol (< 1.0 kcal/mol) of experiment.

We realize that the use of larger more flexible ligands may lead to difficulties in using the straight-forward method suggested here. This is due to the large number

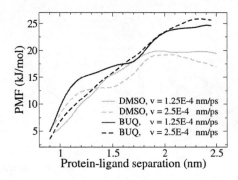

FIGURE 4. The PMF ($\Phi(r)$ in (1) and (4)) as a function of the protein-ligand separation for DMSO (grey) and BUQ (black), shown for both slow (solid) and fast (dashed) speeds. These PMF curves were used to generated the ΔG_{bind} estimates shown in Tab. 1.

TABLE 1. Comparison of computed and experimental binding affinities. All energy values are in units of kJ/mol. The first column describes the ligand used. The second column contains the number of work values N used in the estimate, and the third is the corresponding speed of the spring attached to the ligand. The fourth column shows the binding affinity estimate using Eq. (5). Uncertainty estimates are given in columns five and six, respectively, computed via the bootstrap method, and the approach from Refs. [44, 55]. Finally, the experimental results reported in Ref. [29] are given in the last column. All computational results are within less than 3.0 kJ/mol (< 1.0 kcal/mol) of the experimental data.

Ligand	N	Speed (nm/ps)	ΔG_{bind} Eq. (5)	Uncty (boot)	Uncty (bias [44, 55])	Exp [29]
DMSO	10	1.25×10^{-4}	-11.6	1.6	1.3	-9.7
	20	2.5×10^{-4}	-11.1	2.4	2.0	
BUQ	10	1.25×10^{-4}	-16.7	2.6	1.9	-18.9
	20	2.5×10^{-4}	-18.2	1.2	1.0	

of possible conformations the ligand may adopt in the apo state; all of which must be sampled adequately to obtain accurate PMF curves. However, the method may be modified by including an additional restraint to the RMSD of the ligand, thus restricting the conformational freedom of the ligand. The free energy of release from this RMSD restraint must then be included in the binding affinity estimate [15, 17, 21].

CONCLUSIONS

We have demonstrated that non-equilibrium unbinding simulations utilizing a physical pathway can be used to generate accurate estimates of the binding affinity for the FKBP-DMSO and FKBP-BUQ systems studied here. The computational estimates are in excellent agreement (< 1.0 kcal/mol) with experimental binding data.

The importance of pursuing methods such as described here is that such non-equilibrium approaches are trivially parallelizeable since each unbinding simulation is performed independently. Also, due to the use of a physical pathway, the method is simple to implement in many existing simulation packages with no modification to the software.

We note that the straight-forward approach described here may not produce accurate binding affinities for large, flexible ligands. For such ligands, it will be necessary to extend the approach to include additional restraints to the ligand during the unbinding simulation to prevent large-scale fluctuations. The contribution to the binding affinity from these additional restraints must then be taken into account [15, 17, 21].

The results obtained here suggest that non-equilibrium unbinding simulations can be used to generate accurate estimates of binding affinities. Efficiency analysis and comparison to other methodologies will be carried out in future work.

ACKNOWLEDGMENTS

Funding for this research was provided by the University of Idaho, NSF-EPSCoR and BANTech. Computing resources were provided by IBEST at University of Idaho, and TeraGrid. FMY would like to thank Daniel Zuckerman and Ronald White for helpful discussion.

REFERENCES

1. P. A. Bash, U. C. Singh, F. K. Brown, R. Langridge, and P. A. Kollman, *Science* **235**, 574–575 (1987).
2. J. Hermans, and L. Wang, *J. Am. Chem. Soc.* **119**, 2707–2714 (1997).
3. M. K. Gilson, J. A. Given, B. L. Bush, and J. A. McCammon, *Biophys. J.* **72**, 1047–1069 (1997).
4. W. Chen, C.-E. Chang, and M. K. Gilson, *Biophys. J.* **87**, 3035–3049 (2004).
5. V. Helms, and R. C. Wade, *J. Am. Chem. Soc.* **120**, 2710–2713 (1998).
6. B. C. Oostenbrink, J. W. Pitera, M. M. van Lipzip, J. H. N. Meerman, and W. F. van Gunsteren, *J. Med. Chem.* **43**, 4594–4605 (2000).
7. S. B. Dixit, and C. Chipot, *J. Phys. Chem. A* **105**, 9795–9799 (2001).
8. N. K. Banavali, W. Im, and B. Roux, *J. Chem. Phys.* **117**, 7381–7388 (2002).
9. S. Boresch, F. Tettinger, M. Leitgeb, and M. Karplus, *J. Phys. Chem. B* **107**, 9535–9551 (2003).
10. C. Oostenbrink, and W. F. van Gunsteren, *J. Comput. Chem.* **24**, 1730–1739 (2003).
11. J. M. J. Swanson, R. H. Henchman, and J. A. McCammon, *Biophys. J.* **86**, 67–74 (2004).
12. H. Fujitani, Y. Tanida, M. Ito, G. Jayachandran, C. D. Snow, M. R. Shirts, E. J. Sorin, and V. S. Pande, *J. Chem. Phys.* **123**, 084108–1–084108–5 (2005).
13. D. A. Pearlman, *J. Med. Chem.* **48**, 7796–7807 (2005).
14. J. Carlsson, and J. Aqvist, *J. Phys. Chem. B* **109**, 6448–6456 (2005).
15. H.-J. Woo, and B. Roux, *Proc. Natl. Acad. Sci. USA* **102**, 6825–6830 (2005).
16. Y. Deng, and B. Roux, *J. Chem. Threory Comput.* **2**, 1255–1273 (2006).
17. J. Wang, Y. Deng, and B. Roux, *Biophys. J.* **91**, 2798–2814 (2006).
18. D. L. Mobley, J. D. Chodera, and K. A. Dill, *J. Chem. Phys.* **125**, 084902–1–16 (2006).
19. G. Jayachandran, M. R. Shirts, S. Park, and V. S. Pande, *J. Chem. Phys.* **125**, 084910–1–12 (2006).
20. M. S. Lee, and M. A. Olson, *Biophys. J.* **90**, 864–877 (2006).
21. D. L. Mobley, J. D. Chodera, and K. A. Dill, *J. Chem. Theory. Comput.* **3**, 1231–1235 (2007).
22. J. G. Kirkwood, *J. Chem. Phys.* **3**, 300–313 (1935).
23. R. W. Zwanzig, *J. Chem. Phys.* **22**, 1420–1426 (1954).
24. J. P. Valleau, and D. N. Card, *J. Chem. Phys.* **57**, 5457–5462 (1972).
25. S. Kumar, J. M. Rosenberg, D. Bouzida, R. H. Swendsen, and P. A. Kollman, *J. Comput. Chem.* **13**, 1011–1021 (1992).
26. C. Jarzynski, *Phys. Rev. Lett.* **78**, 2690–2693 (1997).
27. D. Zhang, J. Gullingsrud, and J. A. McCammon, *J. Am. Chem. Soc.* **128**, 3019–3026 (2006).

28. F. Gräter, B. L. de Groot, H. Jiang, and H. Grubmüller, *Structure* **14**, 1567–1576 (2006).
29. P. Burkhard, P. Taylor, and M. D. Walkinshaw, *J. Mol. Biol.* **295**, 953–962 (2000).
30. D. Van Der Spoel, E. Lindahl, B. Hess, G. Groenhof, A. E. Mark, and H. J. C. Berendsen, *J. Comput. Chem.* **26**, 1701–1718 (2005).
31. G. Hummer, and A. Szabo, *Proc. Natl. Acad. Sci. USA* **98**, 3658–3661 (2001).
32. S. Park, F. Khalili-Araghi, E. Tajkhorshid, and K. Schulten, *J. Chem. Phys.* **119**, 3559–3566 (2003).
33. S. Park, and K. Schulten, *J. Chem. Phys.* **120**, 5946–5961 (2004).
34. C. Jarzynski, *Phys. Rev. E* **56**, 5018–5035 (1997).
35. G. E. Crooks, *Phys. Rev. E* **61**, 2361–2366 (2000).
36. C. H. Bennett, *J. Comput. Phys.* **22**, 245–268 (1976).
37. N. Lu, J. K. Singh, and D. A. Kofke, *J. Chem. Phys.* **118**, 2977–2987 (2003).
38. M. R. Shirts, E. Bair, G. Hooker, and V. S. Pande, *Phys. Rev. Lett.* **91**, 140601 (2003).
39. N. Lu, D. A. Kofke, and T. B. Woolf, *J. Comput. Chem.* **25**, 28–40 (2004).
40. N. Lu, D. Wu, T. B. Woolf, and D. A. Kofke, *Phys. Rev. E* **69**, 057702 (2004).
41. M. R. Shirts, and V. S. Pande, *J. Chem. Phys.* **122**, 144107–1–16 (2005).
42. F. M. Ytreberg, S. R. H., and D. M. Zuckerman, *J. Chem. Phys.* **125**, 184114–1–11 (2006).
43. G. Hummer, *J. Chem. Phys.* **114**, 7330–7337 (2001).
44. J. Gore, J. Ritort, and C. Bustamante, *Proc. Natl. Acad. Sci. USA* **100**, 12564–12569 (2003).
45. H. M. Berman, J. Westbrook, Z. Feng, G. Gilliland, T. N. Bhat, H. Weissig, I. N. Shindyalov, and P. E. Bourne, *Nucl. Acids Res.* **28**, 235–242 (2000).
46. A. W. Schuettelkopf, and D. M. F. van Aalten, *Acta Cryst. D* **60**, 1355–1363 (2004).
47. W. F. van Gunsteren, S. R. Billeter, A. A. Eising, P. H. Hünenberger, P. Krüger, A. E. Mark, W. R. P. Scott, and I. G. Tironi, *Biomolecular Simulation: The GROMOS96 manual and user guide*, Hochschulverlag, Zürich, 1996.
48. H. J. C. Berendsen, J. P. M. Postma, W. F. van Gunsteren, and J. Hermans, *Intermolecular Forces*, Reidel, Dordrecht, 1981.
49. W. F. van Gunsteren, H. J. C. Berendsen, and J. A. C. Rullmann, *Mol. Phys.* **44**, 69–95 (1981).
50. H. J. C. Berendsen, J. P. M. Postma, W. F. van Gunsteren, A. DiNola, and J. R. Haak, *J. Chem. Phys.* **81**, 3684–3690 (1984).
51. B. Hess, H. Bekker, H. J. C. Berendsen, and J. G. E. M. Fraaije, *J. Comput. Chem.* **18**, 1463–1472 (1997).
52. T. Darden, D. York, and L. Pedersen, *J. Chem. Phys.* **98**, 10089–10092 (1993).
53. M. P. Allen, and D. J. Tildesley, *Computer Simulation of Liquids*, Oxford University Press, New York, 1989.
54. W. Humphrey, A. Dalke, and K. Schulten, *J. Mol. Graph. Model.* **14**, 33–38 (1996).
55. D. M. Zuckerman, and T. B. Woolf, *Phys. Rev. Lett.* **89**, 180602 (2002).

The Oligomeric Nature of Triosephosphate Isomerase.
Studies of Monomerization

Francisco Zárate-Pérez and Edgar Vázquez-Contreras*.

*Instituto de Química, Departamento de Bioquímica, Universidad Nacional Autónoma de México,
Circuito Exterior, México, DF 04510, México.*
Corresponding autor: E-mail: vazquezc@servidor.unam.mx.

Abstract. In this work, we report the implications of the monomerization of triosephosphate isomerase (TIM) from *T. cruzi* (TcTIM). A monomeric mutant (monoTcTIM) of this species was constructed by genetic engineering, shortening the main loop of interdigitation, which is fundamental for its dimerization. The properties of monoTcTIM were compared with those of the other TIM monomeric versions: genetically engineered mutants, or equilibrium intermediates obtained by chemical denaturation. The stability for almost all the monomeric variants so far reported appears in the same range, and also presents similar structural characteristics. Regarding the catalytic activity of monomeric mutants, when it is present is in several orders of magnitude lower than those observed in their respective wild-type enzyme.

The change in the hydrophobic surface of TcTIM after and before monomerization was also studied and corresponds to an extensive area of the interface region, which becomes exposed when monomers are dissociated. This fact could be related with conformational changes in the local environment of the catalytic amino acids responsible for the isomerization of the substrate and consequently the inactivation of its catalytic properties. Comparison of the structural, folding and unfolding properties, as well as stability studies could give answers on why this enzyme is an obligate oligomer.

Keywords: **triosephosphate isomerase; unfolding; monomeric mutant; monomeric intermediate; monomerization**

INTRODUCTION

Almost ten percent of the most efficient enzymes in nature belong to the superfamily of triosephosphate isomerase (TPI or TIM) barrel proteins (1, 2). TIM barrel enzymes catalyze a variety of reactions, although almost all of them are related to key metabolic pathways. Because of its abundance, the TIM barrel fold is of great interest for protein design as well as for protein evolution research. An extensive literature on the structure and evolutionary relationships of this class of proteins is now available. Regarding protein folding studies with homologous proteins, in the early reports it was reported that the folding pattern is conserved trough evolution; however, in the literature different folding pathways for homologous proteins with the

CP978, *Biological Physics, 3rd Mexican Meeting on Mathematical and Experimental Physics*
edited by L. Dagdug and L. García-Colín Scherer

same denaturant condition have been reported. In this context, TIM is one of the most studied proteins (3). The oligomeric nature of TIM has been analyzed on enzymes mainly from mesophile organisms, where it is always a homodimeric protein, being the simplest model of oligomerization. Also TIM studies have been generated a great interest because the attempts to understand why this protein must have a quaternary structure to be catalytically competent, if the catalytic residues are self-contained in the individual monomers. Notwithstanding, monomeric TIM barrels carrying different catalytic activities do exists in nature. The main function of TIM is to ensure the net production of ATP in the conversion of glucose to pyruvate, and hence, its function is essential for maintaining life under anaerobic conditions. These facts make of TIM a target for drug design against human anaerobic parasites (4, 5). Comparison of amino acid sequence of TIM illustrates their high degree of evolutionary conservation throughout evolution; for example, the sequence around the active site residue (the glutamic acid 168) is totally conserved. The crystallographic structures of wild-type TIMs and engineered mutants have been determined for some species comprising from *Archaea* (6, 7) and *Bacteria* (8-11) to *Eukarya* (12-21). Each TIM subunit has ~250 amino acid residues with a molecular mass close to 27 kDa and folds into a $(\beta/\alpha)_8$ scaffold denominated "TIM barrel" (22). The loops at the carboxyl termini of the barrel form the interface and the active site residues are self-contained in each monomer (23).

TIM barrel structure. The common TIM barrel scaffold contains eight parallel β strands which are Hydrogen bonded to each other forming a barrel-like structure in the core of the protein; this barrel is surrounded by a shield of eight α helices connected to the preceding strands by loop regions of varying length. The α helices cover the β strands in such a way that their hydrophobic face is towards the β sheet and their hydrophilic region interacts with solvent. On the other hand, βα loops (the loops connecting a β strand to the following α helix) tend to be longer than the αβ loops. TIM structure has two protruding loops from the globular structure: the dimer interdigitation loop 3 and the substrate binding loop 6. Most TIMs are homodimers whose quaternary structure is forming when the C-terminal loops (loop 3) of every subunit get buried inside the adjacent monomer during folding (Figure 1a). Two exceptions are the tetrameric TIMs from the thermophilic bacteria *T. maritima* and *P. woesei*, (7, 8).

Folding and stability. All known wild-type TIMs are oligomers. When the enzyme is genetically engineered in order to monomerize it, their catalytic activity is severely diminished. To date, the reason for which wild-type TIMs form at least dimers to be catalytically competent is unknown. It has been postulated that the dimerization would give either the gain in stability upon subunit assembly (24) or that dimerization is necessary for the correct spatial conformation of the active-site pocket (25). In a comparative analysis on the thermostability properties of the enzyme performed by Maes *et al.*, (8) with ten different TIM structures including two thermophilic (*T. maritima* and *Bacillus stearothermophilus*) and one psychrophilic (*Vibrio marinus*), it was found that in both thermophilic and psychrophilic TIMs, an increased amount of salt bridges are present. On the other hand, in the thermophilic enzymes, more hydrophobic surface was found to be buried during folding compared with the other TIMs.

Figure 1. Secondary and tertiary structure of TcTIM and the monomeric mutant from TcTIM. A) Along the barrel axis. The protruding interdigitation loop 3 is visible at the left bottom of the Figure. B) Modeling of the monoTcTIM using the Swiss Model program (http://swissmodelexpasy.org), where the shortened loop 3 can be observed. **(PDB entry 1TCD) (18)**

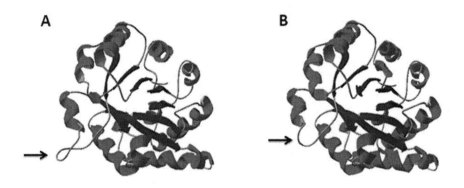

The dimer interface. In TIM, the dimer interface is formed mainly by loops 1-4. Loop 3 protrudes from the monomer and extends into a deep pocket formed by loops 1, 2, and 4 of the adjacent subunit interacting with Lys13 and Glu97 through van der Waals contacts and Hydrogen bonds. For example, a Hydrogen bond is formed between Thr75 in loop 3 and the carboxylate group of Glu97 (26). Two of the active-site residues come from loops 1 and 4 (Lys13 and His95), and the active site is actually located at the dimer interface, although all the residues that directly interacts with the substrate molecule come from the same subunit. The interactions between loop 3 and loops 1 and 4 from every subunit influence the final positioning of active-site residues from loops 1 and 4, respectively. In the dimer interface of *L. mexicana* TIM, a Gln residue present in all other TIMs (27) is replaced by a Glu. In despite of this primary sequence variation, the *L. mexicana* TIM is as stable as the others. Thus, other interactions must be present in the protein to compensate the introduction of an ionizable group (Glu) in the buried interface. Mutating this Glu for Gln (the E65Q mutant) originates a superstable enzyme with similar catalytic properties to those observed in the wild-type enzyme (27). The T_m of this variant is almost 30 °C higher than the obtained for the wild-type protein.

The monomerization of TIM by protein engineering studies. Monomeric TIM variants, in which some regions of the enzyme mainly at the interface were mutated in order to obtain genetic engineering monomeric proteins, have been reported. With these protocols, monomeric variants of TIM have been developed and some of them retain activity while others are inactive. The activity observed for these monomeric mutants, is always many orders of magnitude lower than those observed for the wild-type enzyme. The first genetic engineering monomerization studies of TIM were performed in the enzyme from *T. brucei* (TbTIM), were the fifteen hydrophobic residues of the loop 3 have been changed by 8 more hydrophilic ones (28). The monomeric variant, named MonoTIM, is soluble and stable, although exhibits only about one thousandth of the catalytic activity observed in the dimeric wild-type enzyme (26). Crystallographic studies of these monomeric TIM mutants show that the positions of loops 1 and 4 are severely affected by the shortening of loop 3. The Loop 1 which contains the catalytic residue lysine after mutation, becomes so disordered that it is impossible observe it in the obtained electron density maps.

In order to obtain another monomer from TbTIM, Schliebs and co-workers (25) performed two point mutations on the tip of loop 3 of TbTIM, and the resulting activity of this variant was as low as that observed in the original MonoTIM. Combining modeling with mutagenesis and crystallographic studies, variants of MonoTIM have been designed. In one of these studies, the phosphate-binding loop 8 has been shortened, in order to make the active-site pocket wider and capable of binding substrates other than DHAP and DGAP (33), the resulting protein is stable although inactive. However the crystal structure showed a well-defined conformation of the new loop 8 and a binding widened pocket. Another approach has been followed by Saab-Rincón *et al.* (34) using directed evolution trying to improve the catalytic activity of MonoTIM. It seems that the loss of activity in the monomeric variants is largely due to increased flexibility of the active-site loops, as the stabilizing interactions with the other subunit have been lost. Replacing these intersubunit interactions with intrasubunit ones might lead to recovery of activity. The randomizing strategy was applied to the whole gene sequence in general and also in loop 2; the resulting mutants (A43P and T44A/S) presents an 11-fold increase in kcat when is comparing with the wild-type enzyme. Monomeric variants of human TIM have also been produced after performing point mutations (24, 29). The TIM of *P. falciparum* was also monomerized (30).

In order to go deep in the understanding of the oligomeric nature of TIM, in this work we study the properties of monoTcTIM. This monomeric TIM mutant was thermodynamically characterized in their unfolding and refolding reactions and their structure-stability-function relationships were compared with other monomeric versions of TIM available to date. The relevance of these results is also discussed in order to understand the folding and evolution process in homologous proteins.

Folding and unfolding Studies. Denaturation studies on TIM have been performed widely. Unfolding patterns of the wild-type TIM from different species have been studied using physical and chemical denaturants, namely urea, Gdn-HCl, temperature and pressure. Sometimes the denaturation reaction is reversible, while in others, unspecific irreversible aggregation appears. For some species, the unfolding model is described by a two-state transition (24, 35-38), while in others, more complex

78

processes, including intermediates, have been observed (3, 32, 39-43). These intermediates are responsible for the aggregation and irreversibility observed during unfolding experiments *in vitro*. The intermediates reported in the folding pathway of TbTIM and TcTIM show two states of oligomerization: a dimeric intermediate, which is catalytically competent and less frequently found that the other state, a monomeric intermediate, always inactive (3, 32). Notwithstanding that both enzymes share the same folding pathway induced by Gdn-HCl, TbTIM is irreversible (32) while TcTIM is fully reversible (3).

Design and Expression of a new monomeric TIM: MonoTcTIM. Based on the reversibility difference after equilibrium denaturation in Gdn-HCl between TcTIM and TbTIM (3,32) and in many other characteristics reported for this enzymes (4, 18, 44-57), it was designed, constructed, purified and characterized a monomeric variety of TcTIM. This protein was constructed following the protocol reported for the construction of monoTIM from TbTIM (28). The strategy was performed taking into account other consideration as: 1.- the superposition of the intersubunit contacts of the interface which shows an RMS difference of 0.38 Å for 66 equivalent atoms in ten residues of loop 3 (18). 2.- the different arrangement of the interface Cys from a monomer with the residues of loop 3 of the other monomer would not seem to account for the marked difference in sensitivity of TcTIM and TbTIM to sulfhydryl reagents (46); and 3.- it was reported the generation of heterodimers between these both enzymes after denaturation (57). Considering that TcTIM and TbTIM are very similar at the interface level both in primary sequence and in three-dimensional levels, a genetic construction for a monomeric variant of TcTIM was designed in order to know the structural and catalytic properties of the isolated subunit of this enzyme. Ten equivalent mutations were performed in TcTIM as in the previously reported monomer from TbTIM. The designed protein is identical in sequence to the wild-type TcTIM enzyme except for the loop 3, which is seven residues shorter and has seven substitutions: I69G, G77N, E78A, V79D, S80A, Q82A and I83S. After sequencing, the construction was cloned into an expression plasmid and *E. coli* cells were transformed with the vector and selected by antibiotic resistance. The obtained colonies were grown in liquid medium in order to over-express monoTcTIM. The protein was purified following the procedure reported for the wild-type enzyme (48). The resulting protein was named "monoTcTIM" (Figure 1b), and is a stable monomeric mutant of TcTIM; this variant exhibits significant secondary and tertiary structure, that are very similar respecting the wild-type enzyme as was observed by means of spectroscopic evaluations (circular dichroism and intrinsic fluorescence assays) (Table 1). Additional analyses in order to know the molecular weight of monoTcTIM were performed and showed a molecular weight of 27 kD; with this data the Stokes Radii of the protein were calculated: 22.2 and 20.3 Å were obtained, from Size Exclusion Chromatography (SEC) and Dynamic Light Scattering (DLS) determinations respectively. On the other hand, the calculated stability for this enzyme (ΔG) was 3.1 kCal mol^{-1}.

Table 1. Structural characteristics of monoTcTIM. The percentages for every structural level were obtained from Circular Dichroism and Intrinsic Fluorescence experiments respectively. The molecular weight was obtained by DLS, MALDI-TOF and SEC experiments.

Protein	Molecular Weight	Secondary structure	Tertiary Structure
TcTIM	27 kD (each monomer)	100	100
monoTcTIM	26.4 kD	99.6	99

Comparisons of the obtained molecular weight (MW), Stokes radii (Sr) and stability (ΔG) of MonoTcTIM with other monomeric TIM variants previously reported in the literature showed similar characteristics. Nevertheless a variation in stability was obtained when monoTcTIM is compared with the monomeric intermediate obtained in the equilibrium unfolding pathway of the wild-type enzyme. This difference in stability between the equilibrium intermediate (4.25 kCal mol^{-1}) (3) and the monomeric mutant (3.1 kCal mol^{-1}) can be attributable in some extent, to the method used to calculate the ΔG; for the equilibrium intermediate the obtained value could be overestimated because the unfolding pathway of the wild-type TcTIM in Gdn-HCl is described by a four state process, while in the monomeric mutant the ΔG is consider being more accurate because is described by a two state process. Notwithstanding the aforementioned observations, all of the values obtained in monomeric TIM variants reported so far, shows a well defined range of similar ΔG values (3, 24, 29, 35, 38, 41).

Regarding the size of the TIM monomeric variants expressed as Stokes radii (Sr), two species are observed as compact and expanded conformers. In yTIM where only one equilibrium intermediate has been reported (41), the conformer presents almost the same size of the native homodimer i.e. it is considerably more expanded, respecting the size of the wild-type subunit in the oligomer. On the other hand, in TIM from human and tripanosomatids, the size for the isolated monomer is compact, independently of their genetic engineering or equilibrium intermediate origin; in fact it is very close to the magnitude presented for the monomer in the wild-type dimer (3, 24, 29, 32, 38). Even though, the general structure and the absence of catalysis are common to the monomeric intermediates from different TIM species, their size is variable.

In the early studies of acquisition of three-dimensional structure in proteins it was concluded that the folding pattern will be conserved in homologous proteins, and as discussed before (3) this supposition is not true at all, as there are examples of homologous proteins in which the folding pattern differs, being TIM the most widely studied. The reasons for the selection of TIM from different species to perform folding studies resides in characteristics such as their elevate catalysis, reason why this protein has been named a "perfect catalyst" (1), the abundance on nature of this scaffold (1, 2), and because its oligomeric nature, TIM presents a quaternary structure in despite of the presence of the catalytic residues in the individual monomer. In the origins of folding studies of this enzyme, it was believed that the low stability of the subunits was the main reason for the obligate oligomeric nature of TIM (24, 29). In this sense, the stability of monoTcTIM is now available and the ΔG for the unfolding of the monomeric intermediate in the wild-type enzyme is the highest reported to date (3).

Structural and catalytical properties of the TIM monomer. The most common procedure to monomerize TIM is performing mutations on loop3. The first protocol showed that mutations of the interface loop 3 (28) originates stable monomeric units. It has been also reported other changes including: deletions (33, 58), substitutions (24, 29 25, 30, 38, 59) and combinations of these alterations (34, 53). The change in polarity (25, 30, 38) and in size of the side chains of residues composing the interface (25, 30, 53) were also explored; even though it was found that point mutations produce monomeric proteins (25, 30, 53, 59). In some cases, the monomeric nature of some of these species depends on protein concentrations and exists in equilibrium toward dimerization (34, 59). As was aforementioned, some of these engineered proteins show catalytic activity (26, 30, 34, 53, 58, 59) and others are inactive (24, 29, 33, 38). Some of them have been crystallized and x ray diffracted (33, 38, 58, 60). The position of the mutations in many cases was performed at interface level and in the dimeric wild-type enzyme is mainly composed by loop 3. These interface component interdigits into the pocket of the active site of the other subunit, interacting with residues of loop 1 and 4. The insertion of loop 3 pulls loop 1 from the other subunit into the active site cavity; in TbTIM it was also reported that this introduction forms a salt bridge between residues Lys13 and Glu97 (26). The formation of this bond pulls loop 4 in such a way that His95 is positioned in the correct three-dimensional position to perform catalysis. The arrangement of active site in TcTIM is nearly identical with those of human TIM and TbTIM, for example the RMS differences between the catalytic residues Lys14, His96 and Glu168 in the closed monomers of TcTIM and TbTIM were 0.4, 0.2 and 1.2 Å, respectively. In the open conformation, the corresponding values were 0.3, 0.1 and 1.0 Å (18). Observations in the monomeric engineered TIMs shown that in general no dependence exists with the region mutated at the interface and the resulting high degree of the three-dimensional structure conserved in the obtained proteins. In the case of monoTcTIM it conserves structure at secondary and tertiary levels as has been described in the literature for others interface mutants. However, this elevate amount of general structure is not enough to produce catalysis. On the other hand, when catalysis is finding in some monomeric mutants, it is only in amounts considerably lower to those observed in the wild-type enzyme (26, 30, 34, 53, 58, 59). Then, although the mutation in different sites of the interface produces monomeric TIM barrels, the finest arrangement to perform the perfect catalytic site is not obtained after folding, at least with the regions and substitutions explored up to now.

Another procedure to obtain monomeric varieties of TIM has been explored; it consists in incubating the native homodimer in chemical denaturants. The first report of this monomeric intermediate was obtained at the equilibrium denaturation of yTIM in Gdn-HCl. This conformer appears at mild denaturant concentrations (41, 42, 43). The presence of this intermediate was also observed in the partial denaturation with the same denaturant of TbTIM (32) and in TcTIM (3) (Table 2). The main difference between these enzymes is that in TbTIM, its monomeric intermediate tends to aggregation, turning the unfolding irreversible and difficult to know its stability. On the other hand, in TcTIM the folding pathway is highly reversible, even with the presence of two intermediates (monomeric and dimeric). The monomeric intermediate of TcTIM has a considerable amount of secondary and tertiary structures (3). As observed in some monomeric mutants, this intermediate is inactive and although a method designed to quantify TIM catalytic activity in the presence of denaturant was developed (3, 32), no evidence of activity was found. In order to evaluate if monoTcTIM presents any catalytic activity, a coupled assay enzyme tracking NADH oxidation was used (61). As much as 100 ng mL^{-1} of monoTcTIM as final protein concentration was used in the assay and no catalytic activity was detected. This result suggests that no matter the high equivalence in sequence and three-dimensional arrangement in the two homologous TIMs (TcTIM and TbTIM); it was not possible to obtain catalysis in monoTcTIM performing the same mutations that produce monoTIM from TbTIM. Although the final spatial conformation of the active site in TIM is totally conserved through evolution, the construction of their vicinity is delicate, although flexible, since mutating in different sites in the wild-type interface, produces both monomeric inactive and active mutants.

It has been discussed that the low stability of TIM monomers is neither the only, but not the main, cause of their dimeric nature and that there is an interaction between stability and function implying the appearance of a high ΔG_{assoc} to perform catalysis. It is likely that there is not evolutionary pressure that favors the appearance of stable TIM monomers (43). Although there is a short range for the stability of the monomer, its amount, independently of their genetic engineering or intermediate origin, is always around the fifth part of the total ΔG, and may be for this reason it has not been possible to construct a monomeric variant of triosephosphate isomerase with enough activity levels as those observed in the respective wild-type enzyme. In fact, in the TIM of the hyperthermophilic Archaeas (62), the quaternary structure is represented by a homotetramer, and then through evolution it was primordial to conserve the oligomeric state of TIM. The reason for the low stability of the monomers is possible due to their null possibility to perform enough catalysis. In addition, *in vivo*, the synthesis of TIM is constant and encounters between monomers producing oligomerization, are common. Thus, it is possible that in the cell, the recently formed monomeric intermediate is inactive and unstable because the resulting dimerization enhances activity in several orders of magnitude. On the other hand, efforts to obtain monomeric genetic engineering variants of TIM also produce unstable monomers which present very low or none catalytic activity. These engineered proteins conserve the amount of stability presented in the inactive although highly structured monomeric intermediates produced by partial denaturation. Then, to obtain a monomeric TIM catalytically competent as in the levels observed in the wild-type enzyme is maybe required to increase the stability of the monomer, as has been suggested before (34).

ANS binding. One emerging observation from monomerizing TIM studies is that many hydrophobic wild-type interactions should be lost by the impossibility of oligomerization. Then, the monomeric variants, independently of their equilibrium unfolding or genetic engineering origin, must expose wide interface hydrophobic areas to the solvent. Evidence of this hypothesis was confirmed by the use of ANS. Binding of ANS to predominantly hydrophobic areas of the proteins are accompanied by a large increase in its fluorescence quantum yield (63-66).

The monomeric engineered versions as well as the equilibrium monomers on the unfolded pathways of TIM, binds ANS (3, 32, 41, 43). Dissociation of the TIM monomers originates an exposition to the solvent of all the interface hydrophobic amino acids which originates the increase in fluorescence signal when it is exposed to the dye. The last observation suggests that the formation of the monomers for either the genetic engineering (41, 42) or equilibrium denaturation could be equivalents. Concerning the monoTcTIM, a similar increment in fluorescence by ANS was also found as observed in the equilibrium intermediate previously reported (3) (Table 2).

Table 2. **ANS Fluorescence of the monomeric forms of TcTIM.** Extrinsic fluorescence of a mild unfolding monomeric intermediate and the monoTcTIM were explored. It is showed the fluorescence intensities by ANS (ANS FI) for each protein and their corresponding percentages (%) (see the text for details).

Protein	Condition	ANS FI	% Fluorescence
TcTIM	native	174714	100
monoTcTIM	native	823059	471
TcTIM	1.4 M Gdn-HCl	737434	422

And as just mentioned before, the monomeric intermediate produced by the mild denaturation of yTIM (41, 42), TbTIM (32) and TcTIM (3), are able to bind the fluorescent dye (see Table 2). The effect of ANS on native monoTcTIM was as high as 5 times increased in extrinsic fluorescence comparing with the native wild-type, TcTIM. This increment in fluorescence is due to the exposition of considerable amounts of hydrophobic surface to the solvent in monoTcTIM. This increase is almost in the same amount as that observed in the monomeric intermediate obtained after mild denaturation in Gdn-HCl. Consequently, the significant exposure of non polar surface could be the main reason for the low stability of the monomeric species of TIM, independently of their genetic engineering or mild denaturation origin. For example, the buried surface areas in TcTIM were 1476 and 1491 Å^2 for monomers A and B respectively in the dimer (18), and it is very possible that the amount of this surface increases after mutations in the interface and equilibrium denaturation. On the other hand, in TbTIM, these values were higher than in TcTIM (1522 and 1533 Å^2 respectively) and are maybe the reason of the aggregation behavior that the monomeric intermediate of this enzyme presents under mild denaturant concentrations. Finally, the binding of ANS is commonly used as a characteristic to describe the widely observed folding intermediate denominated "molten globule" (67). It was earlier reported by our group that the monomeric equilibrium unfolding intermediate observed in TcTIM could be a "molten globule" (3), and as was observed for monoTcTIM, both monomers shows considerable amounts of secondary and tertiary structure, they are also non actives and binds ANS in almost the same amount. Hence it is likely that some of the molten globule properties observed in the wild-type monomeric intermediate in Gdn-HCl could also persists in the monomeric mutant. Taking into account that one general objective is to construct a monomeric variant of TIM with enough amounts of catalytic activity as the observed in the wild-type enzyme, and that the low stability of the monomer is possibly the main reason for its inactive or very low active nature, then it is possible that decreasing the hydrophobic exposition after monomerizig could be a solution for recovering the function of the enzyme in a monomeric form.

ACKNOWLEDGMENTS

We thank Laboratorio de Fisicoquímica y Diseño de Proteínas, Facultad de Medicina, Universidad Nacional Autónoma de México, and Dr. A. Gómez-Puyou and Dra. Marietta Tuena, IFC Universidad Nacional Autónoma de México, for generously making available their equipment facilities. We are grateful to Beatriz Aguirre and Ana Lilia Ramírez for their help in purification. This work was supported by Grants 40524M and 41328Q from CONACyT and Grant IN217206 from PAPIIT-UNAM. F.Z.P. is the recipient of a PhD fellowship from CONACyT.

REFERENCES

1. Farber, G. K., and Petsko, G. A. (1990). *Trends Biochem. Sci. 15*, 228-234.
2. Reardon, D., and Farber, G. K. (1995). *FASEB J. 9*, 497-503. 3
3. Chánez-Cárdenas, ME., Pérez-Hernández, G., Sánchez-Rebollar, B.G., Costas, M. and Vázquez-Contreras, E. (2005). *Biochemistry*, 44, 10883-10892.
4. Tellez-Valencia, A., Avila-Rios, S., Perez-Montfort, R., Rodriguez-Romero, A., Tuena de Gomez-Puyou, M., Lopez-Calahorra, F., and Gomez-Puyou, A. (2002). *Biochem. Biophys. Res. Commun. 295*, 958-963.
5. Tellez-Valencia, A., Olivares-Illana, V., Hernandez-Santoyo, A., Perez-Montfort, R., Costas, M., Rodriguez-Romero, A., Lopez-Calahorra, F., Tuena de Gomez-Puyou, M., and Gomez-Puyou, A. (2004). *J. Mol. Biol. 341*, 1355-1365.
6. Walden, H., Bell, G. S., Russell, R. J. M., Siebers, B., Hensel, R., and Taylor, G. L. (2001). *J. Mol. Biol. 306*, 745-757.
7. Walden, H., Taylor, G. L., Lorentzen, E., Pohl, E., Lilie, H., Schramm, A., Knura, T., Stubbe, K., Tjaden, B., Hensel, R. (2004). *J. Mol. Biol.* 342, 861
8. Maes, D., Zeelen, J. P., Thanki, N., Beaucamp, N., Alvarez, M., Dao Thi, M. H., Backmann, J., Martial, J. A., Wyns, L., Jaenicke, R., and Wierenga, R. K. (1999). *Proteins 37*, 441-453.
9. Alvarez, M., Zeelen, J. P., Mainfroid, V., Rentier-Delrue, F., Martial, J. A., Wyns, L., Wierenga, R. K., and Maes, D. (1998). *J. Biol. Chem. 273*, 2199-2206.
10. Noble, M. E. M., Zeelen, J. P., Wierenga, R. K., Mainfroid, V., Goraj, K., Gohimont, A. C., and Martial, J. A. (1993). *Acta Crystallogr. D49*, 403-417.

11. Delboni, L. F., Mande, S. C., Rentier-Delrue, F., Mainfroid, V., Turley, S., Vellieux, F. M. D., Martial, J. A., and Hol, W. G. J. (1995). *Protein Sci. 4*, 2594-2604.
12. Lolis, E., Alber, T., Davenport, R. C., Rose, D., Hartman, F. C., and Petsko, G. A. (1990). *Biochemistry 29*, 6609-6618.
13. Banner, D. W., Bloomer, A. C., Petsko, G. A., Phillips, D. C., Pogson, C. I., Wilson, I. A., Corran, P. H., Furth, A. J., Milman, J. D., Offord, R. E., Priddle, J. D., and Waley, S. G. (1975). *Nature 255*, 609-614.
14. Aparicio, R., Ferreira, S. T., and Polikarpov, I. (2003). *J. Mol. Biol. 334*, 1023-1041.
15. Contreras, C. F., Canales, M. A., Alvarez, A., De Ferrari, G. V., Inestrosa, N. C. (1999). Protein Eng 12, 959-966
16. Wierenga, R. K., Noble, M. E. M., Vriend, G., Nauche, S., and Hol, W. G. J. (1991). *J. Mol. Biol. 220*, 995-1015.
17. Velanker, S. S., Ray, S. S., Gokhale, R. S., Suma, S., Balaram, H., Balaram, P., and Murthy, M. R. N. (1997). *Structure 5*, 751- 761.
18. Maldonado, E., Soriano-García, M., Moreno, A., Cabrera, N., Garza-Ramos, G., Tuena de Gómez-Puyou, M., Gómez-Puyou, A., and Pérez-Montfort, R. (1998). *J. Mol. Biol. 283*, 193-203.
19. Williams, J. C., Zeelen, J. P., Neubauer, G., Vried, G., Backmann, J., Michels, P. A. M., Lambeir, A. M., and Wierenga, R. K. (1999). *Protein Eng. 12*, 243-250.
20. Rodriguez-Romero, A., Hernandez-Santoyo, A., Del Pozo-Yauner, L., Kornhauser, A., and Fernandez-Velasco, D. A. (2002). *J. Mol. Biol. 322*, 669-675
21. Mande, S. C., Mainfroid, V., Kalk, K. H., Goraj, K., Martial, J. A., and Hol, W. G. (1994). *Protein Sci. 3*, 810-821.
22. Knowles, J. R. (1991). *Nature cc20020*, 121-124.
23. Branden, C. I., and Tooze, J. (1991) *Introduction to protein structure*, Garland Publishing Inc., New York.
24. Mainfroid, V., Mande, S.C., Hol., W.G.J., Martial., J., and Goraj, K. (1996). *Biochemistry 35*, 4110-4117.
25. Schliebs. W. Thanki, N., Jaenicke, R. y Wierenga, R.K. (1997). *Biochemistry 36*,9655-9662
26. Schliebs, W., Thanki, N., Eritja, R., and Wierenga R. (1996). *Protein Science*, 5:229-239
27. Williams JC, Zeelen JP, Neubauer G, Vriend G, Backmann J, Michels PA, Lambeir AM & Wierenga RK (1999). *Protein Eng*. 12: 243-250.
28. Borchert, Abagyan, R., Jaenicke, R. y Wierenga, R.K. (1994). *Proc. Natl. Acad. Sci.* 91, 1515-1518
29. Mainfroid, V., Terpstra, P., Beauregard, M., Frère J.M., Mande, S.C., Hol, W.G.J., Martial, J.A., Goraj, K. J. (1996). *J. Mol. Biol.*, 257, 441-456.
30. Maithal, K., Ravindra, G., Nagaraj, G., Singh, S.K., Balaram, H., and Balaram P. (2002). *Protein engineering* 15, 575-584
31. Zomosa-Signoret V, Aguirre-Lopez B, Hernandez-Alcantara G, Perez-Montfort R, de Gomez-Puyou MT, Gomez-Puyou A. (2007). *Proteins*. 67(1):75-83.
32. Chánez-Cárdenas, M. E., Fernández-Velasco, D. A., Vázquez-Contreras, E., Coria, R., Saab-Rincón, G., and Pérez Montfort, R. (2002). *Arch. Biochem. Biophys. 399*, 117-129.
33. Norledge, B.V., Lambeir, A.M., Abagyan, A.R., Rottmann, A., Fernández, A.M., Filimonov, V.V., Peter, M.G., and Wierenga, R.W. (2001). *Proteins: Structure, Function, and Genetics* 42, 383-389.
34. Saab-Rincón, G., Rivelino-Juárez, V., Osuna, J., Sánchez Filiberto y Soberón, J. (2001). *Protein Engineering* 14, 149-155
35. Rietveld, A. W., and Ferreira, S. T. (1996). *Biochemistry cc2002*, 7743-7751.
36. Moreau, V. H., Rietveld, A. W. M., and Ferreira, S. T. (2003). *Biochemistry 42*, 14831-14837.
37. Pan, H., Raza, A. S., and Smith, D. L. (2004). *J. Mol. Biol. 336*, 1251-1263.
38. Lambeir, A. M., Backmann, J., Ruiz-Sanz, J., Filimonov, V., Nielsen, J. E., Kursula, I., Norledge, B. V., and Wierenga, R. K. (2000). *Eur. J. Biochem.* 267, 2516-2524.
39. Beaucamp, N., Hofmann, A., Kellerer, B., and Jaenicke, R. (1997). *Protein Sci.* 6, 2159-2165.
40. Gokhale, R. S., Ray, S. S., Balaram, H., and Balaram, P. (1999). *Biochemistry 38*, 423-431.
41. Vázquez-Contreras, E., Zubillaga, R., Mendoza-Hernández, G., Costas, M., and Fernández-Velasco, D. A. (2000). *Protein Pept. Lett.* 7, 57-64.
42. Morgan, C. J., Wilkins, D. K., Smith, L. J., Kawata, Y., and Dobson, C. M. (2000). *J. Mol. Biol. 300*, 11-16.

85

43. Najera, H., Costas, M., and Fernandez-Velasco, D. A. (2003). *Biochem. J. 370*, 785-792.
44. Zubillaga RA, Perez-Montfort R, Gomez-Puyou A. (1994). *Arch Biochem Biophys*. 1994 Sep; 313(2):328-36.
45. Gomez-Puyou A, Saavedra-Lira E, Becker I, Zubillaga RA, Rojo-Dominguez A, Perez-Montfort R. (1995). *Chem Biol*. 2(12):847-55.
46. Garza-Ramos G, Perez-Montfort R, Rojo-Dominguez A, de Gomez-Puyou MT, Gomez-Puyou A. (1996). *Eur. J. Biochem*. 1;241(1):114-20.
47. Sepulveda-Becerra MA, Ferreira ST, Strasser RJ, Garzon-Rodriguez W, Beltran C, Gomez-Puyou A, Darszon A. (1996). *Biochemistry*. 10;35(49):15915-22.
48. Ostoa-Saloma, P., Garza-Ramos, G., Ramírez, J., Becker, I., Berzunza, I., Landa, A., Gómez-Puyou, A., Tuena de Gómez- Puyou, M., and Pérez-Montfort, R. (1997). *Eur. J. Biochem. 244*, 700-705.
49. Gao XG, Garza-Ramos G, Saavedra-Lira E, Cabrera N, De Gomez-Puyou MT, Perez-Montfort R, Gomez-Puyou A. (1998). *Biochem. J*. 15;332 (Pt 1):91-6.
50. Gao XG, Maldonado E, Perez-Montfort R, Garza-Ramos G, de Gomez-Puyou MT, Gomez-Puyou A, Rodriguez-Romero A. (1999). *Proc. Natl. Acad. Sci*. USA. Aug 31;96(18):10062-7.
51. Perez-Montfort R, Garza-Ramos G, Alcantara GH, Reyes-Vivas H, Gao XG, Maldonado E, de Gomez-Puyou MT, Gomez-Puyou A. (1999). *Biochemistry*. Mar 30;38(13):4114-20.
52. Reyes-Vivas H, Hernandez-Alcantara G, Lopez-Velazquez G, Cabrera N, Perez-Montfort R, de Gomez-Puyou MT, Gomez-Puyou A. (2001). *Biochemistry*. Mar 13;40(10):3134-40.
53. Hernández-Alcantara G, Garza-Ramos G, Hernández GM, Gomez-Puyou A, Perez-Montfort R. (2002). *Biochemistry*.41(13):4230-8
54. Perez-Montfort R, de Gomez-Puyou MT, Gomez-Puyou A. (2002). *Curr Top Med Chem*. May;2(5):457-70. Review.
55. Reyes-Vivas H, Martinez-Martinez E, Mendoza-Hernandez G, Lopez-Velazquez G, Perez-Montfort R, Tuena de Gomez-Puyou M, Gomez-Puyou A. (2002). *Proteins*. Aug 15;48(3):580-90.
56. Tellez-Valencia, A., Olivares-Illana, V., Hernandez-Santoyo, A., Perez-Montfort, R., Costas, M., Rodriguez-Romero, A., Lopez-Calahorra, F., Tuena de Gomez-Puyou, M., and Gomez-Puyou, A. (2004). *J. Mol. Biol. 341*, 1355-1365.
57. Zomosa-Signoret V, Hernandez-Alcantara G, Reyes-Vivas H, Martinez-Martinez E, Garza-Ramos G, Perez-Montfort R, Tuena De Gomez-Puyou M, Gomez-Puyou A. (2003). *Biochemistry*. 42, 3311-8.
58. Thanki, N., Zeelen, J.Ph., Mathieu, M., Jaenicke, R., Abagyan, R.A., Wierenga, R.K., and Schliebs, W. (1997). *Protein engineering*: 10, 159-167
59. Borchert, T.V., Zeelen, J. Ph., Schliebs, W., Callens, M., Minke, W., Jaenicke, R., Wierenga, R.K. (1995). *FEBS Lett.*, *367*, 315-318.
60. Borchert, TV, Prat, K, Zeelen, J.Ph., Callens, M, Noble M.E. M., Opperdoes, F.R., Michels, P.A. M and Wierenga, R.K. (1993). *Eur. J. Biochem*. 211, 703-710
61. Rozacky, E. E., Sawyer, T. H., Barton, R. A., and Gracy, R. W. (1971). *Arch. Biochem. Biophys*. 146, 312-320.
62. Kohlhoff M, Dahm A, and Hensel R. (1996). FEBS Lett. 383(3):245-50.
63. Rosen, C. G., and Weber, G. (1969). *Biochemistry 8*, 3915-3920.
64. Ptitsyn, O. B., Pain, R. H., Semisotnov, G. V., Zerovnik, E., and Razgulyaev, O. I. (1990). *FEBS Lett. 262*, 20-24.
65. Semisotnov, G. V., Rodionova, N. A., Razgulyaev, O. I., Uversky, V. N., Gripas, A. F., and Gilmanshin, R. I. (1991). *Biopolymers 31*, 119-128.
66. Silva, J. L., Silveira, C. F., Correia, A., Jr., and Pontes, L. (1992). *J. Mol. Biol. 223*, 545-555.
67. Ptitsyn, O. B. (1992). The molten globule state, in *Protein Folding* (Creigthon, T. E., Ed.) pp 243-255, Freeman, New York.

Intra-beat Scaling Properties of Cardiac Arrhythmias and Sudden Cardiac Death

Eduardo Rodríguez[a], Claudia Lerma[b], Juan C. Echeverría[a] and Jose Alvarez-Ramirez[a]

[a]*División de Ciencias Básicas e Ingeniería, Universidad Autónoma Metropolitana-Iztapalapa Apartado Postal 55-534, Iztapalapa, D.F., 09340 México*
[b]*Departamento de Instrumentación Electromecánica, Instituto Nacional de Cardiología "Ignacio Chávez", Juan Badiano 1, Col. Sección XVI, Tlalpan, D.F., 14080 México*

Abstract. We applied detrended fluctuation analysis (DFA) to characterize the intra-beat scaling dynamics of electrocardiographic (ECG) recordings from the PhysioNet Sudden Cardiac Death Holter Database. The main finding of this contribution is that, in such recordings involving different types of arrhythmias; the ECG waveform, besides showing a less-random intra-beat dynamics, becomes more regular during bigeminy, ventricular tachycardia (VT) or even atrial fibrillation (AFIB) and ventricular fibrillation (VF) despite the appearance of erratic traces. Thus, notwithstanding that these cardiac rhythm abnormalities are generally considered as irregular and some of them generated by random impulses or wavefronts, the intra-beat scaling properties suggest that regularity dominates the underlying mechanisms of arrhythmias. Among other explanations, this may result from shorted or restricted -less complex- pathways of conduction of the electrical activity within the ventricles.

Keywords: Cardiac arrhythmias; Scaling properties; Complexity.
PACS: 87.19.ug; 05.45.-a; 05.45.Df.

INTRODUCTION

The term cardiovascular physics (CV) has been recently put forward to describe the ongoing interdisciplinary research for gaining deeper insight into pathophysiological conditions of patients suffering from cardiovascular diseases [1]. For instance, the dynamic patterns of cardiac rhythm disturbances may offer clinical information having mechanistic and prognostic implications [2,3]. These disturbances and the interplay with other factors, like transient initiating events and the condition of heart's anatomic and functional substrates, are considered as the main mechanisms leading to sudden cardiac death (SCD) [4]. This health issue continues to be a major prevention challenge as many victims are not known to suffer from heart diseases, thereby considered of having a low risk of dying suddenly [4]. The normal beats (N) arises from the sinus node (situated in right atrium), while a beat that arises from the ventricles is called ventricular ectopic beat (V), premature ventricular complex or PVC (when it happens sooner than expected). Ventricular bigeminy is an arrhythmia identified by a repetitive alternation of a V beats and N beats (e.g. NVNVNVN). More

CP978, *Biological Physics, 3rd Mexican Meeting on Mathematical and Experimental Physics*
edited by L. Dagdug and L. García-Colin Scherer
© 2008 American Institute of Physics 978-0-7354-0497-7/08/$23.00

than 3 consecutive V beats at an abnormally fast rate (more than 100 beats per minute) are called ventricular tachycardia (VT), which can degenerate into ventricular fibrillation (VF) (manifested by disorganized contractions of ventricles that fail to eject blood effectively). The transition from VT to VF, followed by the lack of electrical activity (asystole) appears to be the most common event leading to SCD [5].

Within CV, statistical physics methods are becoming widely applied to analyze cardiovascular data according to temporal, spatial and spatiotemporal behavior. The underlying idea is that, reduced to the most basic terms, cardiac rhythms arise from a small number of fundamental processes involving pacemakers and waves of excitation (depolarization) and recovery (repolarization). The waves can propagate or they can be blocked. If the waves propagate, characteristics of the propagation such as the duration of the action potential and the velocity of the propagation usually depend on the physiological state of the system. All this activity takes place in an anatomically complex structure and is reflected, at least partially, in the ECG record [6]. Several studies have focused on exploring the scaling or fractal properties of the heartbeat intervals fluctuations, showing that abnormal heartbeats interval are less complex than normal ones [7]. Such breakdown of complexity in the heartbeat dynamics has been attributed to the reduction and even loss of feedback control actions (baroreflex) acting in the cardiac dynamics. By recognizing that the ECG is a widely used clinical diagnostic tool, rather than analyzing the heartbeat interval fluctuations, we have investigated the intra-beat scaling properties of ECG recordings from healthy subjects, heart failure patients and patients with episodes of ventricular fibrillation based on detrended fluctuation analysis (DFA). Our results showed that the intra-beat dynamics of healthy subject displays a type of poorly correlated (less regular) behavior, perhaps reflecting the functional integrity of diverse conduction pathways or certain degree of adaptability to changing conditions and the ability of the cardiovascular system to operate within a wide range of operating conditions [8]. On the other hand, in agreement with inter-beat correlation studies, the dynamics of heart failure patients, or even the dynamics manifested during VF, appears to become more regular. In this way, a correspondence between the long term inter-beat and the intra-beat dynamics can be found, so reflecting phenomena at different time scales. That is, in principle, an abnormal heart function can be detected both at short and large time scales, by alterations in the intra-beat and long term inter-beat dynamics respectively, both not necessarily produced by the same underlying mechanisms.

In this work the intra-beat scaling dynamics of ECG recordings involving different types of cardiac rhythm disturbances (arrhythmias) is analyzed. It is found that the ECG waveform becomes more correlated (*i.e.*, less complex) during ventricular bigeminy, ventricular tachycardia (VT), or even atrial fibrillation (AFIB) and ventricular fibrillation (VF) despite the appearance of erratic traces. The results suggest that in these cardiac states, deterministic mechanisms dominate over erratic events. This result may reflect the loss of integrity among diverse conduction pathways of the cardiac tissue or even the reduction of adaptability to changing conditions.

Data

Detrended fluctuation analysis was applied, as described below, to characterize the intra-beat scaling dynamics of 20 electrocardiographic (ECG) recordings from the PhysioNet Sudden Cardiac Death Holter Database [9]. These data were collected from an heterogeneous group of subjects with at least one sustained ventricular tachyarrhythmia leading to cardiac arrest. Subjects aged 35–76 and data were originally sampled at 250 Hz. ECG recordings were also manually scrutinized along time to find and characterize several ECG cardiac abnormalities and arrhythmias. These being considered as prolonged QT interval, flat T wave, ST depression, wide QRS interval, atrial fibrillation (AFIB), PVCs (isolated or during ventricular bigeminy), monomorphic and polymorphic ventricular tachycardia (VT) and ventricular fibrillation (VF).

Detrended Fluctuation Analysis Method

Detrended fluctuation analysis (DFA) is now a widely used method to study fractality and long-term correlations in time sequences. Correlations are measured in terms of a scaling exponent that provides an estimate of the fractal self-similarity dimension of time series fluctuations. In particular, DFA has been used to estimate scaling properties of time sequences of heart rate variability [10-13], being a stable and accurate method that provides a strategy for discriminating between dynamics of healthy and pathological conditions.

In the following, we provide a brief description of this method. For a given time series $y(i), t_i = i\Delta t, i = 1,...,N$ with sampling period Δt the DFA method involves the following steps [14]:

1. Compute the time series mean $\bar{y} = \dfrac{1}{N}\sum_{j=1}^{N} y(j)$ An integrated time series $x(i), 1 = 1,...,N$ is then obtained by:

    $$x(t_i) = \sum_{j=1}^{i} [y(t_j) - \bar{y}], i = 1,.., N$$

2. Divide the integrated time series $x(t_i)$ into boxes of equal size n, which correspond to a time scale $\tau = n\Delta t$. A linear function, denoted by $x_{lin}(t_i;\tau)$, is used to interpolate the sequence in each box. The interpolating curve $x_{lin}(t_i;\tau)$ represents the local trend in each box.

    Compute the fluctuation sequence as

    $$z(t_i;\tau) = x(t_i) - x_{lin}(t_i;\tau), i = 1,...,N$$

3. The fluctuation function $F(\tau)$ is computed as the root-mean squared value of the sequence $z(t_i;\tau)$:

$$F(\tau) = \sqrt{\frac{1}{N}\sum_{j=1}^{N} z(t_j;\tau)^2}$$

4. Repeat the above procedure for a broad range of segment lengths n. According to the recommendations made by Peng et al. [14], a minimum window size $n_{min} \cong 10$ should be selected.

When the signal follows a scaling law, a power-law behavior for the fluctuation function $F(\tau)$ is observed $F(\tau) \approx \tau^\alpha$, where α is called the scaling exponent, a self-affinity parameter representing the long-range power-law correlation properties of the signal. In this way, the scaling exponent α is computed as the slope of the plot $\Im = \{\log(\tau) \ versus \ \log(F(\tau))\}$. In case of having only short-range correlations (or not correlations at all) the detrended walk profile displays properties of a standard random walk (e.g., white noise) with $\alpha = 0.5$. On the other hand, if $\alpha < 0.5$ the correlations in the signal are anti-persistent (i.e., an increment is very likely to be followed by a decrement, and vice versa), and if $\alpha > 0.5$ the correlations in the signal are persistent (i.e., an increment is very likely to be followed by an increment, and vice versa). The values $\alpha = 1.0$ and $\alpha = 1.5$ correspond to $1/f$ noise and Brownian motion, respectively. A value $\alpha > 1.5$ corresponds to long-range correlations that are not necessarily related to stochastic processes. Indeed, $\alpha > 1.5$ can reflect deterministic correlations.

To analyze the data, we proceeded as follows. A preliminary analysis showed that 15 min ECG data sufficed to estimate the intra-beat scaling exponent from the slope of the plot $\Im = \{\log(\tau) \ versus \ \log(F(\tau))\}$ [8]. To retain the ECG intra-beat dynamics, we considered the maximum time scale $\tau_{max} = 0.7s$ for the plot $\Im$. The time evolution of the scaling exponent α was monitored by using 15 minutes overlapping windows, with 80% overlap rate. In this way, we are able to detect scaling exponent changes, and hence correlation properties variations, along the entire ECG record.

RESULTS

Whilst Fig. 1 shows representative traces from five typical sudden cardiac death ECG recordings, Table 1 provides a detailed description of relevant abnormalities and arrhythmias that were found in such recordings. PVCs are labeled as V and the onset of each ventricular tachycardia or fibrillation is indicated by an arrow. The numbers correspond to different relevant events appearing at different time locations along recordings (Table 1). It can be noticed that the ECG complexes do not present a unique behavior. In fact, these are characterized by different waveforms, which can be the consequence of several factors affecting the cardiac muscle activity. From this, it is apparent that identification of VF events by means of direct pattern recognition is quite difficult, involving the determination characterization of a large variety of abnormal waveforms.

Figs. 2 to 4 present the resulting intra-beat scaling dynamics of Table 1 recordings as provided by the behavior of the scaling exponent α along time. Also included in these figures are the corresponding mean heart rate obtained from 15 minutes overlapping windows, percentage of ectopic heart beats and square boxes along the evolution of α showing samples of the ECG trace at the time locations described in Table 1 (circled numbers). The first two examples (Fig. 1, panels A and B) are recordings with normal (sinus) rhythm and very frequent episodes of ventricular

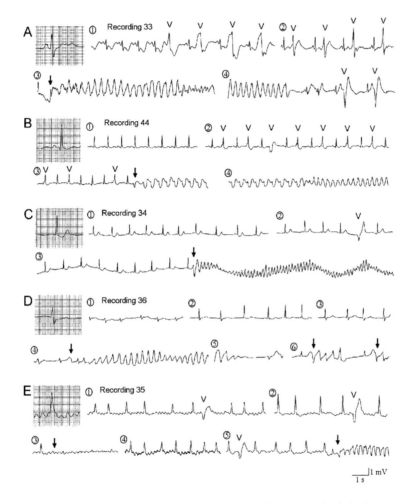

FIGURE 1. Representative electrocardiographic traces from sudden cardiac death database recordings. Each first sample is a heartbeat from the underlying (main) rhythm (grid intervals are 0.04s x 0.1 mV). Premature ventricular complexes are labeled as V and the onset of each ventricular tachycardia or fibrillation is indicated by an arrow. The circled numbers correspond to different time intervals through the recordings (Table 1).

bigeminy (which involved most of the PVCs in both recordings). The increments of ectopic heart beat percent along time in Fig. 2 shows that the intermittent occurrence of the ventricular bigeminy seems associated with higher intra-beat α values. Note that the intra-beat α values became even higher during VT, either polymorphic (case 33) or monomorphic (case 44). In contrast, Fig. 1 panel C shows the recording 34 sinus rhythm interrupted by very few isolated PVCs and a sudden onset of polymorphic VT that converted into VF. Although the fluctuations of the intra-beat α values along time (Fig. 3) does not seem associated to any changes in the ECG features or PVCs, the intra-beat α values also increased before the onset of VT and even more during VT/VF.

In the last two examples (Fig. 1, panel D and E) the underlying cardiac rhythm was AFIB and the ECG shows wide QRS complex. Fig. 4 shows that none of these cases had an association between the occurrence of the PVCs and the intra-beat α values. In fact, case 36 had a very low incidence of PVCs during all recording except at the end where it had a sudden increment of PVCs due mostly to episodes of non-sustained VT.

TABLE 1. Relevant characteristics of the electrocardiographic samples from the sudden cardiac death database.

ID #	Main rhythm	Relevant features	Mark	Time intervals hh:mm:ss	s	Rhythms
A) 33	SR	Prolonged QT interval	1	5:58:57	21537	SR → bigeminy
			2	23:24:20	84260	Bigeminy
			3	15:41:10	56470	Polymorphic VT
			4	15:41:47	56507	VT → bigeminy
B) 44	SR	Flat T wave	1	5:33:20	20000	SR
			2	10:33:21	38001	Bigeminy
			3	19:38:39	70719	PVCs → Monomorphic VT
			4	20:00:24	72024	Monomorphic → polymorphic VT
C) 34	SR	ST depression	1	1:46:55	6415	SR
			2	6:21:25	22885	Isolated PVC
			3	6:35:35	23735	SR → polymorphic VT then VF
D) 36	AFIB	Wide QRS interval	1	5:33:20	20000	AFIB
			2	6:28:47	23327	AFIB → SR with normal conduction
			3	12:30:04	45004	AFIB
			4	18:58:55	68335	AFIB → polymorphic VT
			5	18:59:15	68355	VT → AFIB
			6	19:50:04	71404	Non-sustained VT
E) 35	AFIB	Wide QRS interval	1	4:56:22	17782	AFIB with a PVC
			2	16:39:56	59996	AFIB with a PVC
			3	21:48:39	78519	Monomorphic VT
			4	23:00:42	82842	AFIB with rapid rate
			5	24:34:50	88490	Polymorphic VT

The great oscillations in the intra-beat α values of case 36 could be associated to intermittent episodes of normal intra-ventricular conduction (as is shown in the inset 2), which requires further investigation to be confirmed. Case 35 had both wide QRS and high intra-beat α values along the entire recording. Both recordings had increased intra-beat α values during their VT episodes.

It is observed that the scaling exponent presents higher values for abnormal heartbeats. Also, notice that a relation between the scaling exponent and the mean heart rate only appears to become evident at episodes of VT or VF. Yet the scaling exponent also changes considerably in relation to other types of arrhythmias as revealed in some cases by the ectopic beat percentage (e.g., bigeminy) and by the square boxes accompanying the scaling exponent traces.

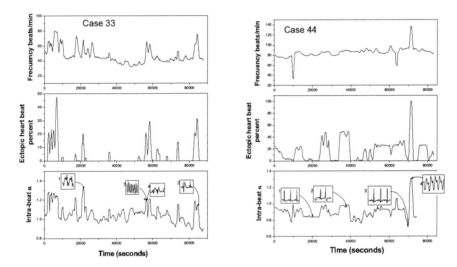

FIGURE 2. Behavior of the scaling exponent α for case 33 and case 44 which present frequent episodes of ventricular bigeminy. Also included the corresponding mean heart rate, the percentage of ectopic heart beats and boxes showing samples of the ECG trace at the time locations described in Table 1 obtained from 15 minutes overlapping windows.

In general, what becomes important to identify in these results is that the existence of arrhythmias generates changes in the intra-beat scaling dynamics towards higher values of alpha, so implying a more correlated (i.e., regular) behavior of the ECG waveform. In particular, in 15 out of 20 recordings analyzed, we found that the intra-beat scaling exponent increases (in some cases abruptly) above unity and reaches a maximum value during episodes of VT and VF (Fig. 2-6). Moreover, in accordance with our previous findings for heart failure patients [8], the intra-beat scaling exponent shows values above those obtained from healthy subjects ($\alpha \cong 0.5$). Particularly in 19 out 20 cases analyzed here the scaling exponent mean values along recordings were found to be near or above the unity ($\alpha_{prom} \geq 0.9$) (Fig. 5).

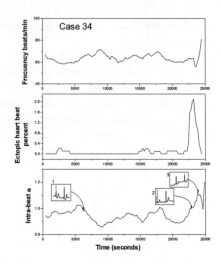

FIGURE 3. Behavior of the scaling exponent α for case 34 which had very few isolated PVCs and a sudden onset of polymorphic VT that converted into VF. Also included the corresponding mean heart rate, the percentage of ectopic heart beats and boxes showing samples of the ECG trace at the time locations described in Table 1 obtained from 15 minutes overlapping windows.

DISCUSSION

Cardiac arrhythmias are produced by different electrophysiological mechanisms. Whereas ventricular ectopic beats and bigeminy arise either from reentrant pathways, ventricular after depolarizations or spontaneous pacemaker [15], fibrillation results from reentry or from multiple areas generating apparently random impulses [16]. Given that the database analyzed here involve heterogeneous group of subjects with different cardiac abnormalities and limited clinical information, it is not possible to clearly know the heart's electromechanical condition of studied cases. Yet the main finding of this work is that the ECG waveform becomes more regular during bigeminy, ventricular tachycardia (VT) or even atrial fibrillation (AFIB) and ventricular fibrillation (VF) despite the appearance of erratic traces. Notwithstanding that these cardiac rhythm abnormalities are generally considered as irregular and some of them generated by random impulses or wavefronts, the intra-beat scaling properties suggest that regularity dominates the underlying mechanisms of arrhythmias.

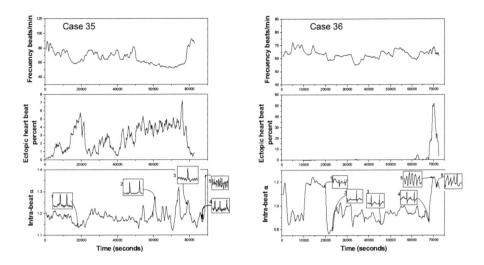

FIGURE 4. Behavior of the scaling exponent α for case 35 and case 36 which present wide QRS complex. Also included the corresponding mean heart rate, the percentage of ectopic heart beats and boxes showing samples of the ECG trace at the time locations described in Table 1 obtained from 15 minutes overlapping windows.

In the database analyzed here it has been identified previously a subgroup of patients that had very frequent and persistent ventricular bigeminy and other set of ECG characteristics, associating this arrhythmia to the mechanism of triggered activity due to early afterdepolarizations [17]. Other types of ventricular arrhythmias showing

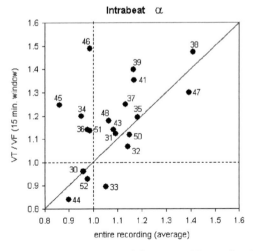

FIGURE 5. Values of intra-beat alphas calculated from 20 sudden cardiac death database recordings (only recordings with at least one episode of VT or VF where selected). The label close to each point indicates the corresponding recording number.

95

different patterns were observed in other recordings from the same database. However, the association between the patterns of the arrhythmias and their mechanism of origin has not been elucidated yet. In this work, the cause of the increment of correlation or ECG regularity under pathological conditions is unclear. However, it could be related to a loss of complexity of the dynamics properties of cellular activation and repolarization of heart tissue. In fact, Kim et al. [18] have shown that a decrease in the number of wave fronts in ventricular fibrillation by tissue mass reduction can cause a transition from chaos to quasi-periodicity. The mechanisms by which the number of wave fronts decreases can be attributed to a reduced heart tissue mass, which in turn reduces the number of invading repolarization wave fronts. When the boundary to mass ratio is increases by e.g. CHF effects, it is more likely for the reentrant wave front to terminate by arriving at a boundary that reduces the complexity of the repolarization dynamics [18]. In fact, this may result from shorted or restricted (i.e., less complex) pathways of conduction. In terms of DFA, this is reflected an increment in the intra-beat scaling exponent α.

It has been conjectured that an increased number of PVCs can be associated with higher risk of sudden cardiac death [19]. In this way, medication oriented to reduce ectopic events would, in principle, reduced cardiac risk. However, clinical trials showed that this is not the case and a reduction of ectopic episodes, via medication, even increased the sudden death risk [20]. Our results have shown that, in many cases, the increment in the number of ectopic beats is accompanied with an increment of the scaling exponent. Apparently the PVCs reduce, on average, the complexity of the heart's conduction pathways. Bigeminy and VT are two arrhythmias with the highest incidence of PVCs. However, ectopic beats are also accompanied with abnormal heartbeats such as bigeminy and polymorphic VT. In this way, according to previous findings [20], it becomes clear that ectopic events are reflecting only a secondary dissipative phenomena, and that the main contribution to the loss of complexity is in the abnormal heartbeat events.

CONCLUSIONS

The vast majority of studies on the complexity of abnormal heart functioning have focused on beat-to-beat interval dynamics. Interesting results have been identified, such as the loss of complexity during abnormal episodes. In a previous work [8], we initiated the study of intra-beat dynamics from raw ECG records, to provide information on the complexity of the heart functioning at scales smaller than a one cycle beat duration. Our results showed that the loss of complexity is more evident for intra-beat than for inter-beat dynamics. In this work, we advanced in the problem by studying potential relations between the loss of complexity and the presence of abnormal heartbeat events and ectopic beats. We found that sudden increments in the scaling exponent, hence a reduction of complexity towards regularity, can be frequently associated to the advent of arrhythmias. For some cases, the reduction of complexity is accompanied by an increment of the number of ectopic beats. However, we did not a clear relation between ectopic episodes and scaling exponent increments.

Further studies are then required incorporating multifractality measurements to elucidate changes in nonlinearity during abnormal heartbeat events.

REFERENCES

1. N. Wessel, J. Kurths, W. Ditto and R. Bauernschmitt, *Chaos* **17**, 15101 (2007).
2. V. Schulte-Frohlinde, Y. Ashkenazy, A.L. Goldberger, PCh. Ivanov, M. Costa, A. Morley-Davies, H. E. Stanley, and L. Glass, *Physical Review E* **66**, 031901 (2002).
3. P.A. Varostos, N.V. Sarlis, E.S. Skordas and M.S. Lazaridow, Appl. Phys. Lett. **91**, 064106 (2007).
4. D. P. Zipes and H. J. J. Wellens, *Circulation* **98**, 2334-2351 (1998).
5. M. Rubart and D.P. Zipes, *J. Clin. Invest.* **115**, 2305-2315 (2005).
6. C. Antzelevitch, *Am. J. Physiol. Heart Circ. Physiol.* **293**, H2024-H2038 (2007).
7. PCh. Ivanov, L.A.N. Amaral, A.L. Goldberger, S. Havlin, M.G. Rosenblum and Z.R. Struzik, *Nature* **399**, 461-465 (1999).
8. E. Rodriguez, J. C. Echeverria and J. Alvarez-Ramirez, *Physica A* **384**, 429-438 (2007).
9. A.L. Goldberger, L.A.N. Amaral, L. Glass, J.M. Hausdorff, PCh. Ivanov, R.G. Mark, J.E. Mietus, G.B. Moody, C. –K. Peng and H.E. Stanley, *Circulation* **101**, e215-e220 (2000).
10. PCh. Ivanov, A. Bunde, L.A.N. Amaral, S. Havlin, J. Fritsch-Yelle and R.M. Baevsky, *Europhys. Lett.* **48**, 694–698 (1999).
11. C.-K. Peng, L. Mietus, J.M. Hausdorf, S. Havlin, H.E. Stanley and A.L. Goldberger, *Phys. Rev. Lett* **70**, 1147–343 (1993).
12. S. Havlin, L.A.N. Amaral, Y. Ashkenazy, A.L. Golberger, PCh. Ivanow and C.-K. Peng, *Physica A* **274**, 99–113 (1993).
13. Y. Ashkenazy, PCh. Ivanov, S. Havlin, C.-K. Peng, A.L. Goldberger and H.E. Stanley, *Phys. Rev. Lett.* **86**, 1900–1910 (2001).
14. C.-K. Peng, S.V. Buldyrev, S. Havlin, M. Simons, H.E. Stanley and A.L. Goldberger, *Physical Review E* **49**, 1685-1689 (1994).
15. P.F. Cranefield. The conduction of the Cardiac Impulse. New York: Futura Publishing Co. 1975.
16. F.H. Samie and J. Jalife. *Cardiovasc. Res.* **50**, 242-250 (2001).
17. C. Lerma, C.F. Lee, L. Glass, A.L. Goldberger. *J Electrocardiol.* **40**, 78-88 (2007).
18. Y.-H. Kim, A. Garfinkel, T. Ikeda, T.-J. Wu, Ch.A. Athill, J.N. Weiss, H.S. Karagueuzian and P.-Sh. Chen, *J. Clin. Invest.* **100**, 2486-2500 (1997).
19. J.T. Bigger Jr, J.L. Fleiss, R. Kleiger, J.P. Miller and L.M. Rolnitzky, *Circulation* **69**, 250-258 (1984).
20. D.S. Echt, P.R. Liebson, B. Mitchell, R.W. Peters, D. Obias-Manno, A.H. Barker, D. Arensberg, A. Baker, L. Friedman, H.L. Greene, M.L. Huther and D.W. Richardson, *N. Engl. J. Med.* **321**, 406-412 (1989).

Evaluation of Aged Garlic Extract Neuroprotective Effect in a Focal Model of Cerebral Ischemia

Penélope Aguilera[1], Perla D. Maldonado[1], Alma Ortiz-Plata[2], Diana Barrera[3] and María Elena Chánez-Cárdenas[1*].

[1]*Laboratorio de Patología Vascular Cerebral, Instituto Nacional de Neurología y Neurocirugía "Manuel Velasco Suárez", México D. F. 14269, México*
E- mail: echanez@bq.unam.mx
[2]*Laboratorio de Patología, Instituto Nacional de Neurología y Neurocirugía "Manuel Velasco Suárez", México D. F. 14269, México*
[3]*Departamento de Farmacología, Facultad de Medicina, Universidad Nacional Autónoma de México, México D. F. 04510, México*

Abstract. The oxidant species generated in cerebral ischemia have been implicated as important mediators of neuronal injury through damage to lipids, DNA, and proteins. Since ischemia as well as reperfusion insults generate oxidative stress, the administration of antioxidants may limit oxidative damage and ameliorate disease progression. The present work shows the transitory neuroprotective effect of the aged garlic extract (AGE) administration (a proposed antioxidant compound) in a middle cerebral artery occlusion (MCAO) model in rats and established its therapeutic window. To determine the optimal time of administration, animal received AGE (1.2 mL/kg) intraperitoneally 30 min before onset of reperfusion (-0.5R), at the beginning of reperfusion (0R), or 1 h after onset of reperfusion (1R). Additional doses were administrated after 1, 2, or 3 h after onset of reperfusion. To establish the therapeutic window of AGE, the infarct area was determined for each treatment after different times of reperfusion. Results show that the administration of AGE at the onset of reperfusion reduced the infarct area by 70% (evaluated after 2 h reperfusion). The therapeutic window of AGE was determined. Repeated doses did not extend the temporal window of protection. A significant reduction in the nitrotyrosine level was observed in the brain tissue subjected to MCAO after AGE treatment at the onset of reperfusion. Data in the present work show that AGE exerts a transitory neuroprotective effect in response to ischemia/reperfusion-induced neuronal injury.

Keywords: Cerebral ischemia, neuroprotection, middle cerebral artery occlusion, antioxidants.

INTRODUCTION

Cerebral ischemia is the result of the transient or permanent reduction of cerebral blood flow with irreversible and fatal damage to brain tissue. It is an important cause of death and disability in industrialized countries and the main cause of hospitalizations among neurological diseases. To date, there are not efficient curative treatments, and in the past two decades important research have been made to understand the biochemical mechanisms involved in brain damage and to develop novel treatments.

The absence of oxygen and glucose in cerebral ischemia triggers serial events which end up in neuronal death, these events are: 1) the loss of neuronal electric activity as a consequence of the disruption of the ATP-dependent processes; 2) the activation of voltage-dependent Ca^{2+} channels and the increase in intracellular Ca^{2+}; 3) the massive release of excitatory amino acids and the consequent increase of their concentration in the extracellular space; 4) the inhibition of protein synthesis; 5) the generation of reactive oxygen species (ROS) and reactive nitrogen species (RNS) leading to oxidative stress; 6) inflammation and 7) apoptosis. The return of blood flow or reperfusion is determinant for the recovery of cell function; however, in some cases, it also has negative side effects such as the generation of oxidative stress.

Oxidative stress is one of the most important events in the ischemia induced damage. It is generated in ischemia and particularly at reperfusion, since free radical production is considerably increased, overwhelming the cellular antioxidant systems. ROS like superoxide anion (O_2^-), hydrogen peroxide (H_2O_2), hydroxyl radical ($\cdot OH$) and RNS such as nitric oxide ($\cdot NO$), peroxynitrite anion ($ONOO^-$), and nitrogen dioxide ($\cdot NO_2$) are normally produced by the cells. These reactive species provoke lipid peroxidation, membrane damage, dysregulation of cellular processes, and DNA damage. The normal generation of free radicals is restricted to no toxic levels by antioxidant molecules such as glutathione, ascorbic acid and α-tocopherol and a number of antioxidant enzymes, including superoxide dismutase, glutathione peroxidase, and catalase. Since ischemia and reperfusion insults generate oxidative stress as response, ROS represent a valuable therapeutic target in stroke and it is conceivable that the administration of antioxidants may limit oxidative damage and ameliorate disease progression [1, 2]. In this work, we used aged garlic extract (AGE) administration in the middle cerebral artery occlusion (MCAO), an in vivo model of cerebral ischemia in rats. AGE is a commercially available odorless garlic preparation widely studied for its high antioxidant activity and its health-protective potential. We determined its protective effect and established the therapeutic time window of AGE on ischemia and reperfusion-induced cerebral injury.

MATERIALS AND METHODS

Animals

Male Wistar rats (280-350 g) were used. Animals had free access to water and commercial rat chow diet. Rats were maintained under constant conditions of temperature, humidity, and lighting (12 h light: dark cycle). All experiments with animals were carried out strictly according to the "National Institutes of Health Guide for the Care and Use of Laboratory Animals", as well as the "Local Guidelines on the Ethic Use of Animals for Experimentation". During the experiments, all efforts were made to minimize animal suffering.

Surgical Procedures

Transient focal cerebral ischemia was produced through middle cerebral artery occlusion (MCAO) as described by Longa et al. (1989) [3]. A 3-0 nylon monofilament was introduced to occlude the origin of the MCA. After 2 h, the filament was removed and the restoration of blood flow (reperfusion) was allowed. Sham-operated rats underwent the same surgical procedures except that the occluding monofilament was not inserted. Animals recovered from anesthesia were sacrificed by decapitation after specific reperfusion times (1, 2, 3, 4, or 24 h).

Behavioral Testing

The evaluation of neurological deficits was performed 30 min before reperfusion and 30 min before sacrifice. The neurological deficit score was associated to a functional damage by the MCAO. Neurological findings were scored on a five-point scale adapted from [3, 4, 5]. First, neurological status of each rat was measured with five tests. Zero was assigned for absence and 1 for presence of impairment in each test. Then, sum of scores obtained on individual test was used to establish the neurological deficit. 1) forelimb flexion; 2) spontaneous motility; 3) grasping reflex; 4) horizontal bar test and 5) contralateral turns [3, 4, 5].

Quantification of the Infarct Area

The infarct area was determined using 2, 3, 5-triphenyltetrazolium chloride (TTC) staining. The whole brains were removed, chilled at -70°C to slightly harden the tissue, and serial coronal sections (2 mm) were cut. The sections were stained with 1% TTC dissolved in phosphate buffer 100 mM, pH 7.4 for 30 min at 37°C in the dark. The slices were washed twice with saline and fixed in 4% paraformaldehyde for 30 min at room temperature. Slices images were digitalized and the area of infarction was measured.

Oxidative Stress Marker

One hundred milliliters of ice-cold isotonic sterile saline with 400 μL of heparin were transcardially perfused followed by 200 mL of 10% buffered formaldehyde/saline, pH 7.4. The brains were removed, postfixed in 10% formalin for 24 h and then immersed in paraffin. Serial 5 μm sections were obtained and used for immunodetection of 3-nitrotyrosine. Endogenous peroxidase was quenched/inhibited with 4.5% hydrogen peroxide in methanol for 1.5 h at room temperature. Non specific adsorption was minimized by leaving the sections in 3% bovine albumin in PBS for 30 min. Sections were incubated overnight with a 1:700 dilution of 3-nitrotyrosine antibody. After extensive washing with PBS, the sections were incubated with a 1:500 dilution of a peroxidase conjugated anti-rabbit IgG antibody during 1 h. Finally, sections were incubated with hydrogen peroxide diaminobenzidine for 1 min, and counterstained with hematoxylin. Cell counting was performed in 5 random fields distributed along the right hemisphere (n = 4). For image analysis the positive cells (in brown) were determined with a computerized image analyzer KS-300 3.0 (Hallbergmoos, Germany). The sections from the studied groups were incubated under the same conditions, so the immunostaining was comparable among the different experimental groups.

Statistical Analysis

Data are expressed as mean ± standard deviation and analyzed by one-way analysis of variance (ANOVA) followed by a post hoc Tukey test. $P<0.05$ was considered statistically significant.

RESULTS

Effect of AGE Administration in MCAo and determination of the AGE Therapeutic Time Window

The effect of AGE administration at ischemia and reperfusion was evaluated. The infarct area was determined after 2 h of ischemia and 2 h of reperfusion (2I/2R), a condition which always resulted in an infarct area of $30.7 \pm 7.3\%$ in the ipsilateral hemisphere as evident from TTC-stained sections (Fig. 1).

The administration of AGE 30 minutes before the onset of reperfusion (-0.5R) produced an infarct area of $26.0 \pm 18.9\%$ in the 2I/2R group. AGE treatment administrated 1 h after reperfusion (1R) reduced the infarct area to $20.9 \pm 9.7\%$. Although both treatments produced reduction in the infarction (15.3 and 31.9%, respectively), statistical significance was not observed ($P<0.05$) (Fig. 1). When AGE was administrated at the beginning of reperfusion (0R), it produced a statistically significant ($P>0.05$, Tukey) reduction of 70% in the infarct (Fig. 1).

The therapeutic time window of AGE was determined after 2 h of MCAO. AGE was administrated at 0R and the infarct area was measured after 3 and 4 h of reperfusion. The infarct area was $24.9 \pm 21.0\%$ after 3 h and $27.6 \pm 21.4\%$ after 4 h of reperfusion. Compared with the corresponding control (2I/2R group, $30.7 \pm 7.3\%$) the infarct area was reduced 18.9% and 10.1% after 3 and 4 h respectively, with no statistical significance ($P<0.05$) (Fig. 2). These results suggest a transitory AGE protection, peaking after 2 h of reperfusion and decreasing 70% the infarct area.

Since 2 h was the maximum protection time observed with a single dose of AGE, a second dose was given after 2 h of reperfusion. The infarct area was $28.0 \pm 6.6\%$ after 4 h of reperfusion. No extension of the therapeutic window ($P<0.05$) was observed. (Fig. 3). A second dose of AGE delivered after 1 h of reperfusion, revealed infarct areas of $13.5 \pm 13.8\%$ and $25.6 \pm 16.8\%$, after 3 and 4 h of reperfusion, respectively (Fig. 3). Although this second dose induced 56% reduction infarct area (at 3 h) and 16.6% (at 4 h), statistical analysis did not reveal a significant difference ($P<0.05$).

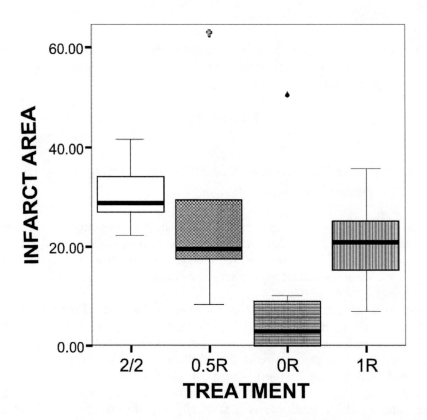

FIGURE 1. The AGE's efficacy depends on the time of administration. Cerebral infarction was induced in rats by middle cerebral artery occlusion during 2 h and rats were sacrificed after 2 h of reperfusion. Rats were treated with AGE (1.2 mL/kg, *i.p.*) at different times: -0.5R, 30 min before reperfusion; 0R, at the beginning of repefusion, and 1R, 1 h after reperfusion. Infarct area was measured in the brain slices by 2, 3, 5-triphenyl tetrazolium chloride staining. The percentages correspond to the average of 3 brain slices of each rat (n=6). Box plots depict the median and the interquartile range between the 25th and 75th percentiles. Whiskers show the range of values that fall within 1.5 times the interquartile range.

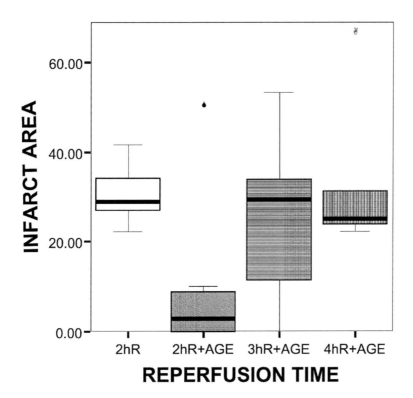

Figure 2. Therapeutic time window of AGE treatment. Cerebral infarction was induced in rats by middle cerebral artery occlusion during 2 h and were sacrificed after different times of reperfusion (2h, 3h and 4 h). Rats were treated with AGE (1.2 mL/kg, *i.p.*) at the beginning of reperfusion. Infarct area was measured in the brain slices by 2, 3, 5-triphenyl tetrazolium chloride staining. The percentages correspond to the average of 3 brain slices of each rat (n=6). Box plots depict the median and the interquartile range between the 25th and 75th percentiles. Whiskers show the range of values that fall within 1.5 times the interquartile range.

Figure 3 shows the effect of forth AGE consecutively doses. The administration of AGE at 0, 1, 2, and 3 h of reperfusion show that after 4 h of reperfusion the infarct area (21.3 ± 7.4%) was reduced (30.6%), but no significant differences were found when compared with the 2I/2R group. (Fig. 3).

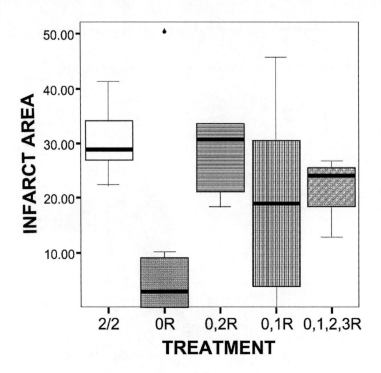

Figure 3. Effect of multiple doses of AGE treatment. Cerebral infarction was induced in rats by middle cerebral artery occlusion during 2 h and were sacrificed after 4h of reperfusion. Rats were treated with AGE (1.2 mL/kg, i.p.) at the beginning of reperfusion and additional doses were given after 1, 2, or 3 h. Infarct area was measured in the brain slices by 2,3,5-triphenyl tetrazolium chloride staining. The percentages correspond to the average of 3 brain slices of each rat (n=6). Box plots depict the median and the interquartile range between the 25th and 75th percentiles. Whiskers show the range of values that fall within 1.5 times the interquartile range.

AGE effect on Neurological Deficit .

Neurological deficit was evaluated with five tests that clearly reveal a dysfunction induced by ischemia in rats (Modified from [3, 4, 5]). None of the animals from control groups (CT, sham, and AGE) showed any motor behavioral abnormalities (Figure 4). Animals subjected to MCAO developed at least 3 of the features upon recovery from anesthesia. Animals from 2I/2R group showed prominent neurological deficits (4.00 ± 0.53) compared with control (P<0.001, Tukey). The neuronal deficit appears during the MCAO and continues during reperfusion (Figure 4).

In animals with decreased infarct damage by the 0R-AGE treatment, a neurological deficit score of 3.50 ± 0.84 was observed. No differences were found when this group was compared to 2I/2R group (P<0.05, Tukey), and no correlation was found between the neurological deficit and the infarct area.

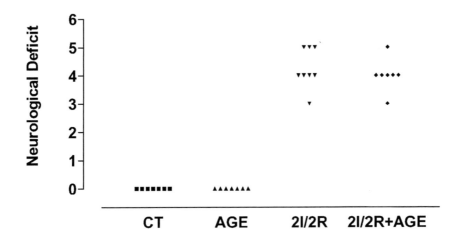

Figure 4. Rats were submitted to 2 h of ischemia and 2 h of reperfusion (2I/2R), and treated with AGE (1.2 mL/kg) at the beginning of reperfusion (0R-AGE). Neurological deficit was scored on a five-point scale. Five individual tests (*forelimb flexion, spontaneous motility, grasping reflex, horizontal bar test,* and *circling behavior*) were carried out 30 min before reperfusion (ischemia 1.5 h) and 30 min before sacrifice (reperfusion 1.5 h). The sum of scores obtained on individual test was used to establish the neurological deficit (0 was assigned to normal and 1 for impaired behavior).

Effect of AGE on Nitrosative Stress

The effect of AGE on the nitration of tyrosine residues in proteins as index of nitrosative stress, was determined. The immunoreactivity of nitrotyrosine residues (counted as positive cells) was increased significantly (P>0.001, Tukey) in ipsilateral hemisphere (2I/2R, 31.5 ± 6.69%) compared to control groups (CT, 1 ± 2.11%; AGE, 0.5 ± 1.58%). On AGE treated group the immunoreactivity decreased in the ipsilateral striatum and cortex (2I/2R+0R-AGE, 7.5 ± 4.25%), showing that AGE (P>0.01, Tukey) reduced oxidative damage.

DISCUSSION

Cerebral ischemia is an important cause of death and disability in industrialized countries. It is the result of the transient or permanent reduction of cerebral blood flow which provokes irreversible and fatal damage to brain tissue. No efficient curative treatments exist to date. Biochemical mechanisms involved in brain damage have been determined to understand the ischemic process.

An important event, consequence of ischemic and reperfusion is oxidative stress. The oxidant species generated in this process have been implicated as important mediators of neuronal injury through damage to lipids, DNA, and proteins, therefore, the use of antioxidant molecules may limit oxidative damage and ameliorate disease progression [1]. In this work we used a commercially available odorless garlic preparation, the aged garlic extract (AGE) an antioxidant compound with health-protective potential [6]. AGE and other several garlic products (e.g. oil garlic, aqueous garlic extract) have proved to be beneficial in cerebral ischemia models [7, 8, 9]. However, the mechanisms of action and the therapeutic time window of AGE are unknown.

AGE Effect on the Infarct Area

To determine the effectiveness of AGE administration throughout ischemia and/or reperfusion we administrated AGE at different times and determined the infarct area in brain sections. All infarct area data were obtained with TTC staining at 2 h of ischemia and 2 h of reperfusion (2I/2R), since this condition always resulted in a prominent infarction.

The administration of AGE to the 2I/2R group, 20 min before reperfusion (-0.5R) show a reduction of 15.3% in the infarct area, and the AGE treatment 1 h after reperfusion (1R) produced a 31.9 % of reduction. We observed that the AGE treatment administrated at the beginning of reperfusion (0R group) produced a statistically significant ($P>0.05$, Tukey) reduction of 70% in the infarct size.

The return of blood flow (reperfusion) is determinant for the recovery of cell function; however, in some cases, it also has negative side effects. It has been reported that there is a peak of ROS production ($489 \pm 330\%$ of control) after 20 min of reperfusion [10], suggesting that the initial moments of reperfusion are critical, and supports the result obtained with AGE treatment at the onset of reperfusion (0R), probably by the prevention of ROS-induced damage.

The reduction of infarct area in the -0.5R and the 1R groups were not statistically significant. However, a partial protection is observed and might be associated to scavenge of ROS produced at ischemia in the -0.5R group. Compared to the 0R group (70% reduction), results obtained with the 1R-AGE treatment suggest that AGE was administrated to late to prevent the oxidative damage induced by the burst of ROS after reperfusion. While -0.5R-AGE

106

treatment was not enough to defend against the high production of ROS during reperfusion since antioxidants components where consumed during ischemia.

Determination of the Therapeutic Time Window of AGE

In order to determine the therapeutic time window of AGE, we administrated the AGE after 2 of MCAO at 0R and evaluated the infarct area at 3 and 4 h of reperfusion. The AGE protection was transitory, with just 18.9% and 10 % respectively reduction in infarct area.

In the results section we showed the effect of additional doses of AGE. The administration of a second dose after 2h of reperfusion (first dose at 0R, second dose 2R) does not show an extension of the protection initially observed.

The administration of a second dose after 1 h of reperfusion showed infarct reductions of 56% (evaluated 3 h after reperfusion) and 16.6% (evaluated 4 h after reperfusion), however statistical analysis did not reveal a significant difference ($P<0.05$). Finally, the effect of four consecutively doses at 0, 1, 2, and 3 h of reperfusion show that after 4 h of reperfusion the infarct area was 30.6%reduced, but no significant differences were found when compared with the 2I/2R group ($P<0.05$).

This finding suggests that AGE only delay MCAO induced damage but it can not prevent it and that probably the massive production of ROS seen during reperfusion may be scavenged by antioxidants present on AGE.

Effect of AGE on Neurological Deficit

Rats were subjected to five tests to evaluate the neurological deficit. As expected CT, sham, and AGE did not show any motor behavioral abnormalities [Fig. 4]. In contrast, all animals subjected to MCAO developed at least 3 of the features upon recovery from anesthesia. We observed that the neuronal deficit appears during the MCAO and continues during reperfusion. However, the animals with a considerable decrease in the infarct damage (0R-AGE group) show a high neurological deficit score (3.50 ± 0.84), with no differences when compared with the 2I/2R group ($P<0.05$, Tukey).

We suggest that the no correlation between the reduced infarct after 0R-AGE treatment and the high functional impairment may be related to the transitory protector effect of AGE observed in the present study. AGE protection is crucial to diminish the infarct area (representative of the neuronal death), however ischemia damage is observed by the high score in neurological deficit.

Effect of AGE on Protein Oxidative Damage

Nitration of tyrosine residues in proteins is used as and index of nitrosative stress. The immunoreactivity of nitrotyrosine residues was increased significantly in ipsilateral hemisphere compared to control groups and the AGE treatment considerably decreased the number of immunopositive cells in the ipsilateral striatum and cortex suggesting a reduced oxidative damage compared with control. The chemical complexity of AGE makes difficult to understand the mechanisms involved in AGE-mediated protection. The most abundant organosulfur compound on AGE (0.62 mg/g in commercial preparation) [11] is SAC (S-allylcisteine). In vitro, SAC is able to scavenge O2·- and ONOO- [12]. The reduced 3-nitrotyrosine signal observed in Figure 3, suggest that the ONOO- scavenging ability the main or other components of AGE may contribute to the protective effect of AGE after cerebral ischemia observed in this work.

These findings suggest that AGE could be potentially useful in patients with cerebral ischemia subjected to thrombolytic therapy. Nevertheless since the AGE therapeutic window was short, a complementary treatment must be administrated to obtain and extend the protective effect.

REFERENCES

1. P. Aguilera P, M. E. Chánez-Cárdenas, P. D. Maldonado. "Recent advances in the use of antioxidant treatments in cerebral ischemia", in *New Perspectives on Brain Cell Damage, Neurodegeneration and Neuroprotective Strategies* edited by A. Santamaría and M. E. Jiménez-Capdeville, Kerala, India: Research Signpost, 2007. pp. 145-159.
2. I. Margaill, M. Plotkine and D. Lerouet, *Free Radic Biol Med.* **39**, 429-43 (2005).
3. E. Zea-Longa, P. R. Weinstein, S. Carlson, R. Cummins, *Stroke* **20**, 84-91 (1989).
4. S. A. Menzies, J. T. Hoff and A.L. Betz, *Neurosurgery* **31**, 100-6 (1992).
5. M. Modo, R. P. Stroemer, E. Tang, T. Veizovic, P. Sowniski and H. Hodges, *J Neurosci Methods.* **104**, 99-109 (2000).
6. H. Amagase, *J Nutr* **136**, (3 Suppl) 716S-725S (2006).
7. R. Gupta, M. Singh and A. Sharma. *Pharmacol Res.***48**, 209-15 (2003).
8. Y. Numagami, S. Sato and S. T. Ohnishi. *Neurochem Int.* **29**, 135-43 (1996).
9. S. Saleem, M. Ahmad, A. S. Ahmad, S. Yousuf, M. A. Ansari, M. B. Khan, T. Ishrat and F. Islam. *J Med Food.* **9**, 537-44 (2006).
10. O. Peters, T. Back, U. Lindauer, C. Busch, D. Megow, J. Dreier, U. Dirnagl. *J Cereb Blood Flow Metab.* **18**,196-205 (1998).
11. L. D. Lawson, "The composition and chemistry of garlic cloves and processed garlic", in *Garlic: The Science and Therapeutic Application of Allium sativum L. and Related Species*, edited by H. P. Koch and L. D. Lawson, Baltimore: Williams & Wilkins, 1996, pp. 37–107.
12. O. N. Medina-Campos, D. Barrera, S. Segoviano-Murillo, D. Rocha, P. D. Maldonado, N. Mendoza-Patino, J. Pedraza-Chaverri, *Food Chem Toxicol.* **45**, 2030-2039 (2007).

Protein's unfolding and the glass transition: a common thermodynamic signature.

L. Olivares-Quiroz* and L.S. Garcia-Colin†

*Colegio de Ciencia y Tecnologia. Universidad Autonoma de la Ciudad de Mexico. Av La Corona S/N, C.P. 07160, Mexico D.F.
†Departamento de Fisica. Universidad Autonoma Metropolitana-Iztapalapa. Av Michoacan y Purisima. C.P. 09340, Mexico D.F.

Abstract. Recently, it has been recognized that protein's folding and unfolding mechanisms exhibit a wide range of common features with the glass transition observed in supercooled organic and inorganic liquids. Such similarities range from pure thermodynamic aspects such an anomalous ΔC_p and a substantial entropy decrease $\Delta S < 0$, to strictly kinetic aspects as the existence of an excess of vibrational modes at low frequencies (*bosonic peak*) revealed by Raman and neutron scattering experiments. In this work, we discuss the experimental and theoretical facts that might enable an extrapolation of the Adam-Gibbs scheme for the standard glass transition to describe the relaxation time τ as function of temperature T in biological macromolecules' unfolding.

Keywords: glass transition, protein unfolding thermodynamics, Adam-Gibbs theory, averaged relaxation time τ, protein unfolding kinetics
PACS: 81.05.Kf; 61.43.Fs; 87.14.E; 87.15.ad; 87.15.hm; 87.15.hp

PROFILE OF THE GLASS TRANSITION.

The term glass transition usually refers to the supercooling of a liquid below its crystallization temperature T_c without the onset of the standard crystalline structure [1],[2]. Once in the glassy state, the system displays a molecular structure very similar to the previous liquid state in addition to solid-like responses to shear and strain, like a substantially increase of the relaxation time to equilibrium [3], together with an exponential increase of viscosity η as a function on temperature T [4]. From a pure thermodynamic point of view, the glass transition has two characteristic hallmarks; an abrupt decrease of the specific heat C_v in the vicinity of the glass transition temperature T_g and also a decreasing entropy change $\Delta S < 0$ due to conformational molecular rearrangements [5], [6]. During the glass transition, the specific heat C_v decreases even up to a 50 % of the vibrational specific heat C_v in the liquid state [2]. For many authors, this is considered as the most distinctive feature of the glass transition since it is correlated to the breaking of ergodicity and it exhibits the strongly non-equilibrium character of the glassy state [6], [7]. Even when the standard glass transition refers to the existence of non-equilibrium states in inorganic liquids when supercooled, the existence of a glassy state has been also observed and carefully characterized for organic liquids like propylene-glycol [8], monohydroxy alcohols [9]; for ionic mixtures like $Ca_xK_y(NO3)_z$ [10], polymeric structures as poly-methyl methacrylate (PMMA) [11] and more recently for biological macromolecules, from RNA strands [12], to proteins like cytochrome-c [13], serum albumin [14], β-lactoglobulin [15], gelatin [16], among many others. An interesting

CP978, Biological Physics, 3rd Mexican Meeting on Mathematical and Experimental Physics
edited by L. Dagdug and L. García-Colín Scherer
© 2008 American Institute of Physics 978-0-7354-0497-7/08/$23.00

fact regarding the glass transition observed for biological macromolecules is that water itself exhibits a well-characterized glass transition around $T_g = 135$ Kelvin, even though the system composed by proteins and water show a glassy dynamics for temperatures close to $T_g = 190$ Kelvin [17]. One of the most recent explanations for this phenomena suggests that protein glass transition is driven properly by a dynamic transition in the structural water, which induces increased fluctuations in protein's side chains located on the external surface of the protein [18].

According to the standard picture of the glass transition in liquids, the supercooling of a liquid brings a considerable reduction of the volume accessible to the molecules to diffuse with a corresponding increase in the system's density [1]. In such scenario, the probability p of interaction between two or more particles increases significantly and diffusion along the system begins to slow down. The more likely scenario to produce an increasing diffusion time is the formation of clusters of particles where the interaction rate has increased substantially. In a rough sense, these clusters act as kinetic "cages" preventing the system reaching its equilibrium configuration since particles spend most of their time in collisional events rather than in diffusive mechanisms [19], [20]. As temperature decreases, the system can be depicted then by a set of such "cages" in which molecules display mainly vibrational movements around an average position and occasionally move between adjacent clusters [21], [22], [23]. The mechanism described above lies at the center of the most accepted theories for the glass transition, from the semi-phenomenological approach proposed by Adam and Gibbs [24], the Random First Order Transition Theory [19] and the different versions of the Mode Coupling Theory [25], [26], [27]. All such theories have their advantages and shortages in describing the phenomenology associated with the glass transition crossover: the exponential dynamics of viscosity η as a function of temperature T, the change in slope of the Debye-Waller factor near the T_g temperature and the existence of at least two different dielectric relaxation processes in non-conducting glasses (α and β relaxation mechanisms) [5], [22].

PROTEIN'S UNFOLDING SCENARIO.

As previously mentioned, hydrogen-bonded biological polypeptide chains in aqueous solutions also display a glassy state when supercooled far below the standard water's crystallization temperature $T_c = 273$ Kelvin [18], [28]. Particularly, the change of slope of the Debye-Waller factor and the presence of an excess of vibrational modes (bosonic peak) around $T_g \sim [190 - 220]$ Kelvin is well documented for proteins like β-lactoglobulin [29], lysozyme [30], and myoglobin [31], pointing out that far below $T_c = 273$ Kelvin, such biological systems enter in a glassy regime. Albeit the thermodynamic and kinetic description of the glass transition in proteins is by far one of the most active research lines in protein science field, this will not be the main issue addressed in this Letter. Instead, we will focus our attention to protein's native state unfolding driven by temperature. Such type of denaturation has proven to exhibit common thermodynamic characteristics with the standard glass transition in liquids, specifically in what concerns to C_v or C_p and configurational entropy ΔS_c dynamics as functions of temperature T. [1]

From the early calorimetric measurements of the specific heat C_p in small globular proteins it has been clear that protein's denaturation is accompanied by an extensive heat absorption which generates an strongly increase of the isobaric specific heat C_p [32], [33], [34]. The temperature T_m where C_p exhibits a maximum depends on denaturation conditions, namely, pH solution, as well as protein's molecular weight M_w. If the specific heat C_p increase is sharp enough, as it usually occurs for small molecular weight proteins ($M_w < 20$ kDa), the calorimetric enthalpy change ΔH during denaturation can be well accounted by the van't Hoff's unfolding enthalpy ΔH_{vH} [35]. Several explanations have been proposed to account for the significant increase in C_p during denaturation. However, the experimental evidence suggests that the hydrophobic effect [36], namely, the reorganization of the molecular structure of surrounding water due to the exposure of the non polar side chains to the solvent, is responsible for the major part of the heat absorption [37], [38], [39], [40]. In most inorganic supercooled liquids, the glass transition is associated with a negative specific heat change ΔC_v whose steepness is a measure of the fragility or strongness of the glass-forming liquid [1], [4] and it can be considered as constant in the vicinity of T_g. In contrast, protein's unfolding exhibit a $\Delta C_p > 0$ and it is generally non-constant over usual physiological temperature range [35], [34]. In fact, for proteins like chymotrypsin [41], myoglobin and Ribonuclease-A [42], calorimetric measurements for C_p suggest it can be represented by a quadratic polynomial on T.

In addition to a positive ΔC_p in protein's unfolding and a negative ΔC_v for liquids during the glass transition, both phenomena are also characterized in thermodynamic terms by a substantial configurational entropy decrease $\Delta S < 0$. Although the scheme proposed by Adam-Gibbs [24] suggests that the configurational entropy decrease $\Delta S < 0$ during the liquid-glass transition can be associated with the exponential growth of the size of cooperative regions and therefore with an also exponential decrease of the number of accessible configurations to the system [19], the actual reason for an entropy decrease in protein's denaturation is quite different. Protein's folding into a compact globular structure is driven mainly by the hydrophobic collapse of the non-polar residues of the polypeptide chain into the core of the native state [37], [43], [44], [45]. When unfolded, the entropy change of the system arises from two contributions. In the first place, the passage from an ordered structure to an almost random linear polypeptide chain generates a positive $\Delta S_{chain} > 0$. Simultaneously and even more fundamental, it is the fact that the unfolding of the native state exposes the hydrophobic core to the surrounding water molecules causing a self-avoidance interaction between non-polar residues and polar solvent molecules. Since hydrogen-bonding among water molecules is energetically higher than spatial rearrangements, solvent molecules will prefer to rearrange themselves into more ordered structures [39]. This reduction of the configurational possibilities for the solvent induces an even higher entropy decrease $\Delta S_{solv} < 0$

---

[1] Since the most significant contribution to the difference between C_p and C_v is a partial derivative $(\partial p / \partial T)$, $C_p \simeq C_v$ upon folding or unfolding of biological macromolecules.

than the corresponding entropy increase $\Delta S_{chain} > 0$ coming from polypeptide chain's disorder. Around physiological temperatures, the latter effect is significantly larger than the former, and it is finally responsible for the total configurational entropy decrease during protein's unfolding [39].

Thus, according to the phenomenology discussed above, both the standard glass transition in supercooled liquids and small protein's denaturation have common thermodynamic signatures. It is worth to underline we do not refer here to the *actual* glass transition observed for hydrated proteins around $T_g \sim 210$ Kelvin. Our disgression here only points out that protein's native state unfolding, which occurs at a completely different temperature range, shares at least two common properties with the liquid-glass transition: an abrupt change in the corresponding specific heat C_v or C_p together with an entropy decrease $\Delta S < 0$. Given such common thermodynamic background, we go a step further and suggest it is plausible to export the formalism of Adam-Gibbs to calculate the averaged relaxation time τ upon denaturation of the native state. As mentioned before, the Adam-Gibbs scheme for the glass transition proposes that the exponential increase of the relaxation time τ in glasses is driven by a configurational entropy term. To export such formalism to protein's denaturation, we must calculate two entropic contributions. In a first place, the global entropy change ΔS^{config} due to the hydrophobic effect when the polypeptide chain is immersed in a polar solvent, and in the second place, the entropy change Δs^* due to local conformational transitions inherent to the secondary structure formation. In a work already in progress, we establish precisely how to accomplish such calculations for a highly helical protein like myoglobin (PDB: 1MBN), with the use of calorimetric data on myoglobin's unfolding and a differential geometry-based approach to describe the conformational transitions in secondary structure.

ACKNOWLEDGMENTS

One of the authors, L. Olivares-Quiroz, would like to express his gratitude to Dr. Orlando Guzman and Dr. Leonardo Dagdug at Universidad Autonoma Metropolitana-Iztapalapa, and to Dr. Rafael Barrio at Universidad Nacional Autonoma de Mexico for helpful discussions.

REFERENCES

1. Angell. C.A. Formation of glasses form liquids and biopolymers. *Science*, **267**, 1924-1935, (1995).
2. Angell. C.A. The old problems of glass and the glass transition, and the many new twists. *Proc. Natl. Acad. Sci. USA*, **92**, 6675-6682, (1995).
3. Donth E.J. The glass transition: relaxation dynamics in liquids and disordered materials. Springer, Berlin (2001).
4. Debenedetti P.G., Stillinger F.H. Supercooled liquids and the glass transition. *Nature*, **410**, 259-267, (2001).
5. Wales, D.J. Energy landscapes with applications to clusters, biomloecules and glasses. Cambridge University Press. First Edition. (2003).

6. Bouchaud, J.P.,Biroli, G. On the Adam-Gibbs-Kirkpatrcik-Thirumalai-Wolynes scenario for the viscosity increase in glasses. *J. Chem. Phys.*, **121**, 7347, (2004).

7. Monasson, R. Structural glass transition and the entropy of the metastable states. *Phys. Rev. Lett.*, **75**, 2847-2850, (1995).

8. Leheny, R.L., Menon, N., Nagel, S.R., Price D.L., Suyuza, K., and Thirigarayan P. Structural studies of an organic liquid through the glass transition. *J. Chem. Phys.*, **105**, 7783-7794, (1996).

9. Wang, L.M., Richert, R. Glass transition dynamics and boiling temperatures of molecular liquids and their isomers. *J. Chem. Phys B*, **111**, 3201-3207, (2007).

10. Mezei, F., Knaak, W., Farago, B. Neutron spin echo study of dynamic correlations near the liquid glass-transition. *Phys. Rev. Lett.*, **58**, 571-574, (1987).

11. Roth, C.B., Pound A., Kamp S.W., Murray C. A., and Dutcher J.R. Molecular weight dependance of the glass transition temperature of freely standing poly(methyl-methacrylate) films.*Eur. Phys. J. E Soft Matter*, **20**, 441-448, (2006).

12. David F., Wiese K.J., Systematic field theory of the RNA glass transition. *Phys. Rev. Lett.*, **98**, 128102, (2007).

13. Yadaiah M., Kumar R., Bhuyan A.K. Glassy dynamics in the folding landscape of cytochrome-c detected by laser photolysis. *Biochemistry*, **46**, 2545-2551, (2007).

14. Goddard Y.A., Korb J.P., and Bryant R.G. Structural and dynamical examination of the low-temperature glass transition in serum-albumin. *Biophys J.*, **91**, 3841-3847, (2006).

15. Parker R., Noel T.R., Brownsey G.J., Laos K., and Ring S.G. The non-equilibrium phase and glass transition behaviour in β-lactoglobulin. *Biophys J.*, **89**, 1227-1236, (2005).

16. Kasapis S., and Sablani S.S. A fundamental approach for the estimation of the mechanical glass transition temperature in gelatin. *Int. J. Biol. Macromol.*, **36**, 71-78, (2005).

17. Kumar P., Yan Z., Mazza M.G., Buldyrev S.V., Chen S.H., Sastry S., and Stanley H.E. Glass transition in biomolecules and the liquid-liquid critical point of water. *Phys. Rev. Lett.*, **97**, 177802, (2006).

18. Tournier A.L., Xu J., and Smith J.C. Translational hydration water dynamics drives protein glass transition. *Biophys. J.*, **85**, 1871-1875, (2003).

19. Lubchenko V., and Wolynes P.G. Theory of aging in structural glasses. *J. Chem. Phys.*, **121**, 2852-2865, (2004).

20. Lubchenko V., and Wolynes P.G. Theory of structural glasses and supercooled liquids. *Annu. Rev. Phys. Chem.*, **58**, 235-266, (2007).

21. Garcia-Colin L.S., and Goldstein-Menache P. La Fisica de los Procesos Irreversibles. Tomo II. El Colegio Nacional. México. (2003).

22. Garcia-Colin L.S. Remarks on the glass transition. *Rev. Mex. Fis*, **1**, 11-17, (1999).

23. Mezard M., and Parisi G. Statistical physics of structural glasses. *J. Phys. Condens. Matter*, **12**, 665-667, (2000).

24. Adam G., and Gibbs J. On the temperature dependence of cooperative relaxation properties in glass-forming liquids. *J. Chem. Phys.*, **43**, 139, (1965).

25. Sjogren L., and Gotze W. α-relaxation spectra in supercooled liquids. *J. Non-Cryst. Solids*, **172**, 7-15, (1994).

26. Gotze W., and Sjogren L. Comments on the mode coupling theory for structural relaxation. *J. Chem. Phys.*, **212**, 47-59, (1996).

27. Yatsenko G., Schweizer K.S. Ideal glass transition, shear modulus, activated dynamics and yielding in fluids of non-spherical objects. *J. Chem. Phys.*, **126**, 014505, (2007).

28. Teeter M.M., Yamano A., and Mohany U. On the nature of a glassy state of matter in a hydrated protein and dynamics: X-ray crystallographic studies of the protein ribonuclease A at nine different temperatures from 98 to 320 K. *Biochemistry*, **31**, 2469-2481, (2001).

29. Orecchini A., Paciaroni A., Bizzarri A.R., and Cannistraro S. Low-frequency vibrational anomalies in β-lactoglobulin: contribution of different hydrogen classes revealed by inelastic neutron scattering. *J. Chem. Phys. B*, **105**, 12150-12156, (2001).

30. Caliskan G.A, Kisliuk A., and Sokolov A.P. Dynamic transition in lysozyme: role of a solvent. *J. of Non-Crystalline Solids*, **307**, 868-873, (2002).

31. Leyser H., Doster W., and Diehl M. Far-infrared emission by boson peak vibrations ina globular protein. *Phys. Rev. Lett.*, **82**, 2987, (1999).

113

32. Privalov P.L., Khechinashvilii N.N. A thermodynamic approach to the problem of stabilization of globular protein structure: a calorimetric study. *J. Mol. Biol.*, **86**, 665-684, (1974).

33. Gomez J., Hilser V.J. Xie, D., and Freire, E. The heat capacity of proteins. *Proteins*, **22**, 404-412, (1995).

34. Prabhu N.B., and Sharp K.A. Heat capacity in proteins. *Annu. Rev. Phys. Chem.*, **56**, 521-548, (2005).

35. Privalov P.L., and Dragan A.I. Microcalorimetry of biological macromolecules. *Biophys. Chem.* , **126**, 16-24, (2007).

36. Kyte J. The basis of the hydrophobic effect. *Biophys. Chem.* , **100**, 193-203, (2003).

37. Kauzmann, W. Some factors in the interpretation of protein denaturation. *Adv. Protein Chem.* , **14**, 1-63, (1959).

38. Schellman J.A. Temperature, stability and the hydrophobic interaction. *Biophys. J.* , **73**, 2960-2964, (1997).

39. Finkelstein A.V., and Ptitsyn O.B. Protein Physics. Academic Press Elsevier Science USA. (2002).

40. Kumar S., Jung C., and Nussinov, R. Temperature range of thermodynamic stability for the native state of reversible two-state proteins. *Biochemistry*, **42**, 4864-4873, (2003).

41. Brandts J.F. *Biochemistry*, **42**, 4864-4873, (2003).

42. Privalov P.L., and Makhatadze G.I. Heat capacity of proteins I. Partial molar heat capacity of individual aminoacid residues in aqueous solutions. *J. Mol. Biol.*, **215**, 585-591, (1990).

43. Dill K.A., Chan H.S. From Levinthal to pathways to funnels. *Nature Struct. Biol.*, **4**, 10, (1997).

44. Sadqui M., Lapidous L.J., and Munoz V. How fast is hydrophobic collapse?. *Proc. Natl. Acad. Sci. USA*, **100**, 12117-12122, (2003).

45. Walther, M. Signatures of hydrophobic collapse in extended proteins captured with force spectroscopy. *Proc. Natl. Acad. Sci. USA*, **104**, 7916-7921, (2007).

Spatio-temporal dynamics of a three interacting species mathematical model inspired in physics

Faustino Sánchez-Garduño and Víctor F. Breña-Medina

Departamento de Matemáticas, Facultad de Ciencias,
Universidad Nacional Autónoma de México (UNAM),
Circuito Exterior, Ciudad Universitaria, México, 04510, D.F.

Abstract. In this paper we study both, analytically and numerically, the spatio-temporal dynamics of a three interacting species mathematical model. The populations take the form of pollinators, a plant and herbivores; the model consists of three nonlinear reaction-diffusion-advection equations. In view of considering the full model, as a previous step we firstly analyze a mutualistic interaction (pollinator-plant), later on a predator-prey (plant-herbivore) interaction model is studied and finally, we consider the full model. In all cases, the purely temporal dynamics is given; meanwhile for the spatio-temporal dynamics, we use numerical simulations, corresponding to those parameter values for which we obtain interesting temporal dynamics.

Keywords: pollinator-plant-herbivore models, reaction-diffusion-advection equations, Holling functional response of type II.
PACS: 01.30.Cc

INTRODUCTION

The heterogeneity is one of most obvious feature of the environment. In fact, humidity, temperature, food and water distribution, vegetation, local weather, etc., are different from one location to another. Among others, the drastic changes in the weather originates migratory movements of biological populations, like the monarch butterfly from Canada and USA to the central West parts of Mexico during the winter season. The population migratory movements are not at random, they exhibe a spatial coherence which manifests it during the journeys. Thus, depending on the specific individuals, they form: swarms, herds, flocks, schools, etc.

Changing the space scales, the gregarious or social behaviour of some species is on the basis of the grouping. With this, they defend themselves from other species, overcome hostile environmental conditions or for mating. The grouping of microscopical individuals also has been studied: The ameba *Dictyostelium discoideum* exhibe spiral aggregative patterns towards an attractant substance (cAMP). The colonies of the *Bacillus subtilis* bacteria adopt different morphologies —included the branched patterns of fractal type— depending on the agar and nutrient concentrations in the Petri dish.

In all the above described processess the underlying interplay is between the individual and the collective movement. In fact, the grouping processes involves different space and time scales. In this context, the answer to the question: How do the individuals interact each other in order to produce such a collective ordered spatial distributions?

In other side, but closely related, the pattern formation approach covers a wide range of areas and spatial scales, including: fetal develompment, coats of mammals, pigmen-

CP978, *Biological Physics, 3rd Mexican Meeting on Mathematical and Experimental Physics*
edited by L. Dagdug and L. García-Colín Scherer
© 2008 American Institute of Physics 978-0-7354-0497-7/08/$23.00

tation of shells, waves of activation in cardiac muscle, structure of social insects nests, collective swarms of bacteria, army ants, vegetation distributions in arid and semi-arid zones, etc. The crucial thing here is the searching of the physical, chemical or biological processess underlying the emergence the above mentioned ordered structures. Given that some of those can be expressed in mathematical terms, depending on the space and time scales the above mentioned leads to different interesting mathematical problems.

From a mathematical point of view, for this *self-organization* process, there are two main not excludent approaches:

- **Complex systems.** The formation of different grouping patterns is the result of a complex nonlinear cooperative individual behaviour (or communication) — including short distances signaling, sounds, movements, etc.— In this approach the attention is on the individuals with certain coupling within them.
  In [3] the authors present a study for the formation of different type of selforganization processes in vertebrate animals.
- **Continuous.** This look at the bulk, the movement of the population as a whole and establishes laws of movement with an analogy reasoning approach consisting in viewing the population movement as a continuous media like. An alternative way of deriving these equations is by using a random walk approach. One of the precursor of this approach in ecological contexts was Skellam (see [16] and [17]). Since then more work has been carried out by other authors (see [10], [12], [14] and [20]).
  The equations can be local, like partial differential equations (reaction(r)-diffusion(d)-advection(a)) or nonlocal namely integro-differential equations. The derivation of the first, according with the continuous approach, can be done by using two points of view: the Lagrangian and the Eulerian (see [10]).
  At certain space scales of description, the rda equations could be appropriate. In [6], [10], [12] and [14] the authors present, mainly focusing on different type of reaction-diffusion equations, a review of this approach in ecological contexts.

We should say that fundamental ecological processes, like dispersion (see [6]) and [10]) segregation ([5]) and aggregation (see [15] and [19]) of interacting population lead to very important mathematical problems. Of course, the tools to being used to elucidate them, depends on the particular mathematical modelling approach.

In this paper we explore the existence of ordered spatial distributions of some population densities on a rectangular habitat, appearing in three nonlinear partial differential equation systems which describe the following interactions: pollinator-plant, herbivore-plant and pollinator-plant-herbivore, respectiveley. Those are consistent with the ecological coexistence of the interacting populations.

The paper is organized as follows. The second section deals with the temporal and spatio-temporal dynamics of an extreme particular case where we have just a pollinator-plant interaction. In the next section a herbivore-plant interaction is considered. The fourth section contains our study of the full pollinator-plant-herbivore interaction mathematical model. The paper ends in section five where we state what we consider are some open questions in these type of models. Here, we also carried out a final discussion as well.

ONE BASIC EXTREME CASE: THE POLLINATOR-PLANT DYNAMICS

The analysis we present in this section is carried out in two steps. Firstly we consider the temporal dynamics of a mutualistic interaction model which, for interpretation purposes, takes the form of a pollinator and plant populations. Secondly, we incorporate the space component in the previous model and carry out some numerical simulations as well.

The assumptions and the model

It is accepted that the mutualistic interactions are ubiquitous in nature. In spite of that, it is an unattended ecological relationship. In fact, when one compares the amount of theoretical studies on predation or competence with those on mutualism, one realizes that the last one are just a few. In [1] the reader can find —up to the authors knowledge— the best review on mutualism.

The underlying hypothesis of the model we consider here, are: 1. The pollinators, in addition of the nectar and pollen from the plants, they have other limited source, 2. The plants are pollinized exclusively by this pollinator population i.e., they are highly specialized and 3. The pollinator-plant interaction is described by a *Holling functional response of type II*. Qualitatively speaking, this is like the Michaelis-Menten expression for the production rate of the product appearing in enzyme kinetics. In the present context it reflects one fact: the pollinator rate of visits to plants divided by the pollinator population density, is limited i.e., does not grow without any bound. It must be a monotonic growing function for low plant population density but for big enough plant population density, such function has an asymptotic behaviour towards a horizontal straight line representing the maximum rate of visits *per* pollinator.

Let $a(t)$ and $p(t)$ be the population density of pollinator and plants at time t, respectively. One nonlinear model which captures the above hypothesis is

$$
\begin{aligned}
\dot{a} &= a(K-a) + \frac{ap}{1+p} \\
\dot{p} &= -\frac{p}{2} + \frac{ap}{1+p},
\end{aligned} \tag{1}
$$

where the dot on a and p denotes the temporal derivative of these variables. Here K is the parameter bifurcation of the system which, in addition to being the carry capacity for the pollinators popullation, as we are going to see, also represents a measure of the preference of the pollinators for their own food sources. As far we know, the model (1) was originally proposed by Soberón and Martínez del Río (see [18]). These authors did it on the basis of feasible ecological interpretation of all the parameters and expressions appearing in their original formulation. Here, we just kept the very fundamental parameter: K. In the reference [18], the authors only carried out a preliminar and uncomplete analysis. They focussed mainly in discussing the ecological meaning of the model.

The temporal dynamics

In order to obtain the dynamics associated with the system (1), we use the standard nonlinear dyamical tools. The relative position of the nontrivial null-clines

$$p_1(a) = \frac{a-K}{K+1-a} \quad \text{and} \quad p_2(a) = 2a-1,$$

of the system (1), determinates its nontrivial[1] equilibria in the first quadrant of the phase space and the specific form in which they touch each other, gives the local phase portrait of (1).

The values of a for which $p_1(a) = p_2(a)$ are

$$a_1, a_2 = \frac{1}{2}\left[(K+1) \pm \sqrt{(K+1)^2 - 2}\right].$$

We set $K^* = \sqrt{2} - 1$ and $K_1 = 0.5$. In terms of these parameter values, we state the following theorem.

Theorem 1. *Any solution of the system (1) starting from (a_0, p_0) with $a_0 \geq 0$ and $p_0 \geq 0$ is bounded and nonnegative for all t. Moreover (1) has:*

- *not closed trajectories,*
- *only two equilibria: $P_0 = (0,0)$ and $P_1 = (K,0)$, if $0 < K < K^*$. P_1 is global attractor on $\mathbf{R}_+^2$, i.e., the plants population becomes extinct,*
- *three equilibria: P_0, P_1 and $P^* = (a^*, p^*) = (1/\sqrt{2}, \sqrt{2} - 1)$, whenever $K = K^*$. P^* is not hyperbolic of saddle-node type and, depending on both population densities: or the species coexist in equilibrium, or the plants tend to the extinction,*
- *four equilibria: P_0, P_1 and the other two come from the bifurcation of P^*, if $K^* < K < K_1$. Depending on the population densities, the species coexist or the plants become extinct,*
- *three equilibria: P_0, P_1 and P_r if $K \geq K_1$. Here the species coexist by exhibing a global attractor: P_r.*

Part of the proof of the above theorem can be found somewhere else (see [9]). Figures 1-3 show the phase portrait[2] of the system (1) for the relevant parameter values.

The spatio-temporal model and the simulations

Now we are going to incorporate the space component. The hypothesis for the spatio-temporal model are: 1. The pollinators population movement —at individual level— is at random. The meaning of this assumption, at level of population, is: the whole

---

[1] By nontrivial equilibrium, we mean that equilibrium point whose none of its coordinates, are zero.
[2] All the phase portrait figures we present in this paper, were done by using the free *pplane* software.

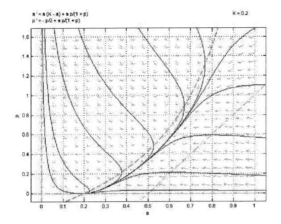

FIGURE 1. Phase portrait of the system (1) for $K = 0.2$

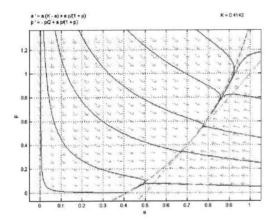

FIGURE 2. Phase portrait of the system (1) for $K = 0.4142$

population moves towards minus the gradient of the population density direction, 2. The plants do not disperse but its spatial distribution changes because of the interaction with the pollinator popullation and 3. The temporal dynamics is given by the system (1).

One model built up on the basis of the previous assumptions is:

$$
\begin{aligned}
\frac{\partial a}{\partial t} &= D\nabla^2 a + a(K-a) + \frac{ap}{1+p} \\
\frac{\partial p}{\partial t} &= -\frac{1}{2}p + \frac{ap}{1+p},
\end{aligned}
\tag{2}
$$

here we are abusing with the notation since $a(\vec{r},t)$ and $p(\vec{r},t)$ denote (the same as in system (1)) the respective population densities at point $\vec{r}$ of the habitat at time t, $D > 0$ is the diffusivity of the pollinator population and ∇^2 denotes the laplacian operator. In order to obtain the numerical solutions of the nonlinear system (2), we use a rectangular,

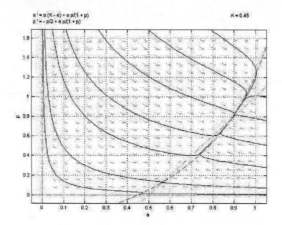

$a' = a(K - a) + a p/(1 + p)$
$p' = -p/2 + a p/(1 + p)$

$K = 0.45$

FIGURE 3. Phase portrait of the system (1) for $K = 0.45$

$\Omega \subset \mathbf{R}^2$, habitat with initial conditions corresponding to a spatial perturbation of a steady and homogeneous state of the system (see the item four of the theorem 1 and figure 1c)) (2) and homogeneous Neumann boundary conditions. The results[3] are shown in figures 4-6. The color scale is as follows: The red represents the highest concentration and purpure the lowest one. In between, appear yellow, green and blue which correspond to descendent values of the concentration.

In this case, the steady and homogeneous state of (2) given by the nontrivial equilibrium of the system (1) acts as an attractor of the possible state of the system i.e., both population species tend to the homogeneous spatial distribution as time goes to infinity. In other words: for the specific initial and boundary value problem associated with the system (2), the diffusion term does not destroys the attractive feature of the homogeneous and stationary state corresponding to an equilibrium P of the system (1).

OTHER EXTREME CASE: PLANT-HERBIVORE INTERACTION

As we mentioned in the Introduction, one of the most studied ecological interaction is predation. Here we consider one interaction which, from a demographic perspective, can be seen as one of predator-prey type. Once again, for interpretation purposes this takes the form of a plant-herbivore interaction.

---

[3] All the numerical solutions of the PDE systems we present in this paper, were done by using the software *FlexPDE* which solves the system by using the finite element method.

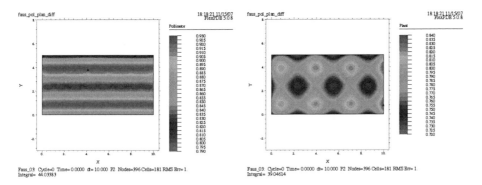

FIGURE 4. Numerical solutions of the system (2): the initial conditions.

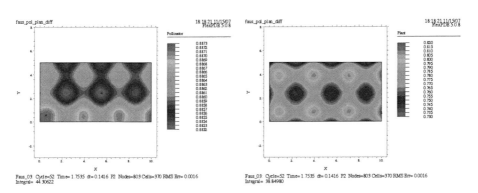

FIGURE 5. A shootcute at runing time $t = 1.57$.

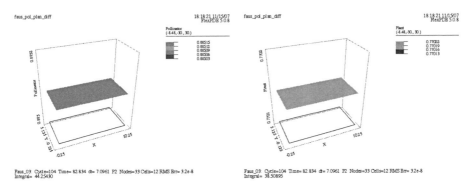

FIGURE 6. The final homogeneous spatial distribution.

The assumptions and the model

We consider the following hypothesis: 1. The plants have limited sources. The meaning of this is: if there are not herbivores, its population density increases logistically. 2. The plants are the unique source of food for the herbivores, i.e., if there are no plants, the herbivores population tends to the extinction, and 3. The plant-herbivore interaction is described by a functional response of type II.

Let us denote by $h(t)$ the population density of the herbivores at time t, then one mathematical model which incorporates the above hypothesis is:

$$\begin{aligned} \dot{p} &= p(1-p/K) - \frac{ph}{(1+p)} \\ \dot{h} &= -\alpha\beta h + \beta\frac{ph}{(1+p)}, \end{aligned} \tag{3}$$

where β and K are positive parameters.

The temporal dynamics

The main branch of the null-clines of the system (3) are

$$h(p) = (1+p)(1-p/K) \quad \text{and} \quad p = \alpha/(1-\alpha),$$

with $0 < \alpha < 1$, which intersect each other at

$$P_0 = (0,0), \quad P_1 = (K,0) \quad \text{and} \quad P_2 = (\bar{p},\bar{h}),$$

where

$$\bar{p} = \frac{\alpha}{1-\alpha} \quad \text{and} \quad \bar{h} = (1+\bar{p})(1-\bar{p}/K).$$

We set

$$p^* = \frac{K-1}{2} \quad \text{and} \quad \tilde{K}(\alpha) = \frac{\alpha+1}{1-\alpha}.$$

The real part of the eigenvalues of the Jacobian matrix of (3) at P_2 is

$$Rea(\lambda_1,\lambda_2) = \alpha h'(\bar{p}),$$

which, given that $0 < \alpha < 1$, changes its sign depending on $h'(\bar{p})$. We now state the following theorem, which summarizes the dynamics associated with the system (3).

Theorem 2.

1. *If $K \leq \frac{\alpha}{(1-\alpha)}$, the herbivore population becomes extinct and that of the plants stabilizes at P_1. This point can be a stable node or a saddle-node,*

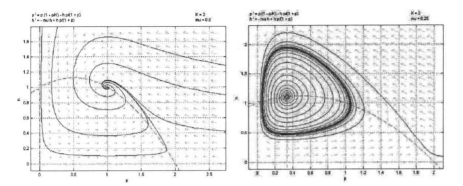

FIGURE 7. Dynamics of the system (3) for different signs of $\alpha h'(\bar{p})$: Negative (left): The equilibrium is an attractor. Positive (right): The emergence of a limit cycle from a Hopf bifurcation.

2. *If $K > \frac{\alpha}{(1-\alpha)}$, P_1 is a saddle point. Moreover:*
   - *If $K < \tilde{K}(\alpha)$, P_2 is asympotically stable and both species coexist through a global attractor in $\mathbf{R}^2_+$. One possibility is through damped oscillations,*
   - *If $K \geq \tilde{K}(\alpha)$, P_2 is unstable and the populations coexist. The population densities tend to an isolated periodic behaviour i.e., to a stable* limit cycle *which surrounds P_2. The limit cycle emerges from a* Hopf bifurcation.

Part of the proof of this theorem can be seen in [8] (Example 3.1). Figure 7 shows the phase portrait of the system (3) and the ecological interpretation of each one, is given in the Theorem 2 statement.

The spatio-temporal model and simulations

As in the previous case, we now are going to include the spatial distribution factors. For this purpose we consider the following assumptions: 1. The spatial distribution of plants only changes due to the interaction with the herbivores, 2. At individual level, the herbivore population moves at random and 3. The temporal dynamics is described by a system of the form (3). The resulting mathematical model is:

$$
\begin{aligned}
\frac{\partial p}{\partial t} &= p(1-p/K) - \frac{m_1 hp}{(s+p)} \\
\frac{\partial h}{\partial t} &= D\nabla^2 h + \frac{m_2 hp}{(s+p)} - \eta h,
\end{aligned}
\tag{4}
$$

where K, m_1, m_2, s and η are positive parameters. For the numerical simulations of the initial and boundary value conditions problem associated with (4), we take the parameter values for which the homogeneous system (3) has a limit cycle emerging from a Hopf bifurcation. The initial conditions we choose are

$$
p(\vec{r},0) = R(1+0.2\sin(x+y)\cos(x-y)) \text{ and } h(\vec{r},0) = W(1+0.4\sin(2x)), \, \forall \vec{r} \in \Omega,
$$

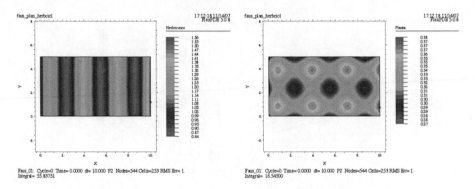

FIGURE 8. Numerical solutions of the system (4) for the parameter values $D = 6$, $K = 2$, $m_1 = 1$, $m_2 = 1$, $s = 1$ and $\eta = 0.25$. Initial conditions.

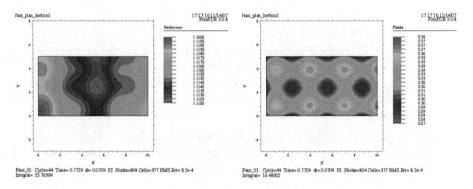

FIGURE 9. Spatial distribution at $t = 0.77$

where $R = 0.364$, $W = 1.11$ which are about ten per cent perturbation of the homogeneous steady state of the system (4) surrounded by the limit cycle. The result of our numerical simulations are shown in figures 8-10. From this, we can see that after some interesting transients including the fact that the herbivore population takes the qualitative initial distribution of the plants i.e., the herbivores coexist with the plants in the same habitat. But for big enough time, both population densities evolve towards spotted temporal periodic oscillations. This is to say, for big enough time, for each fixed point of the rectangular habitat, the local population densities change periodically in time. This resembles us the so-called *breathing patterns* which already have been reported in other contexts. See figure 4.

We should say that in addition to the previous numerical simulations, we also carried out ones for the parameter values for which the equilibrium P_2 is globally asymptotically stable. What we got was, as is was suspected, a final homogeneous spatial distributions in both population densities for big enough time.

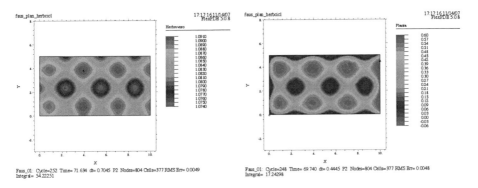

Herbivoro

1.0910
1.0900
1.0890
1.0880
1.0870
1.0860
1.0850
1.0840
1.0830
1.0820
1.0810
1.0800
1.0790
1.0780
1.0770
1.0760
1.0750
1.0740

Planta

0.60
0.57
0.54
0.51
0.48
0.45
0.42
0.39
0.36
0.33
0.30
0.27
0.24
0.21
0.18
0.15
0.12
0.09
0.06
0.03
0.00
-0.03
-0.06

Faus_01: Cycle=252 Time= 71.634 dt= 0.7045 P2 Nodes=804 Cells=377 RMS Err= 0.0049
Integral= 54.22251

Faus_01: Cycle=248 Time= 69.740 dt= 0.4445 P2 Nodes=804 Cells=377 RMS Err= 0.0048
Integral= 17.24298

FIGURE 10. Breathnig final pattern at $t = 69.7$

THE FULL POLLINATOR-PLANT-HERBIVORE MODEL

In this section we consider the full model which describes the interaction of three populations under the form of pollinator-plant-herbivore. To begin with, we should say that one important issue here is to is to investigate the role played by a third species in a mutualistic interaction keept by two populations. One case of this is considered here, where a herbivore species is added to the mutualistic interaction sustained by plants and pollinators.

The assumptions and the model

Here are the hypothesis for the homogeneous model: 1. The pollinators, in addition of having some benefits (nectar) from the plants, also have other limited source of food, 2. The plants are pollinized exclusively by this pollinator population i.e., the plants are hihgly specialized as we mentioned it previously, 3. The pollinator-plant interaction is described by a Holling response of type II, 4. The plants are the unique food source for the herbivores, 5. The plant-herbivore interaction is described by a Holling response of type II, 6. The herbivores interact with the pollinators in an indirect form: by reducing the pollinator visits rate to plants.

A mathematical model buildt up on the basis of the above mentioned hypothesis is

$$
\begin{aligned}
\dot{a} &= ba(K-a) + \frac{g(h)k_2\sigma\mu ap}{1+\sigma\phi\mu^2 p} \\
\dot{p} &= -\gamma p + \frac{g(h)k_1\sigma\mu ap}{1+\sigma\phi\mu^2 p} - \frac{m_1 ph}{c_1+p} \\
\dot{h} &= -\delta h + \frac{m_2 ph}{c_1+p},
\end{aligned}
\tag{5}
$$

where $g \in C^1[0,\infty)$, $g(0) = 1$, $g'(h) \leq 0$ and $g(h) > 0 \; \forall \; h \geq 0$ is the *reduction rate of visits* of pollinators to plants due to the herbivores interaction. All the parameters appearing in (5) are positive and have an important ecological interpretation. In fact: k_1

is the number of fertilized ovums in each pollinator visit, σ is the probability of visits, ϕ is a measure of the speed of nectar extraction and μ is the energetic recompense.

On the temporal dynamics

The points $P_0 = (0,0,0)$ and $P_1 = (K,0,0)$ are equilibria of the system (5) for any parameter values. The main branch of the null-clines of (5) are

$$
\begin{aligned}
f_1 &\equiv b(K-a) + \frac{g(h)k_2\sigma\mu p}{1+\sigma\phi\mu^2 p} &= 0 \\
f_2 &\equiv -\gamma + \frac{g(h)k_1\sigma\mu a}{1+\sigma\phi\mu^2 p} - \frac{m_1 h}{c_1+p} &= 0 \\
f_3 &\equiv -\delta + \frac{m_2 p}{c_1+p} &= 0.
\end{aligned}
\tag{6}
$$

The analysis we present here will be done by considering two main cases depending on the election of the function g.

Case 1. $g(h) \equiv 1$. Here there exists the possibility of an equilibrium in the first positive octant for (5): $\bar{P} = (\bar{a}, \bar{p}, \bar{h})$ where

$$
\begin{aligned}
\bar{a} &= K + \frac{k_2\mu^2\sigma\bar{p}}{b(1+\phi\sigma\mu^2\bar{p})} \\
\bar{p} &= \frac{c_1\delta}{m_2-\delta} \\
\bar{h} &= [(c_1+\bar{p})k_1\mu\sigma\bar{a}]/[m_1\delta(1+\phi\gamma\mu^2\bar{p})],
\end{aligned}
$$

In the following points, we summarize (from [7]) what is known so far on the local phase space of the system (5):

1. P_0 is a saddle point for all parameter values. The two dimensional stable manifold lies in the ph plane. Therefore, if there is pollinator population and the plant and herbivore populations are small enough, both species become extinct,

2. P_1,
   - is a local attractor if $\mu < \gamma/k_1\sigma$. Hence, if the population densities of the three species are in a small neigbourhood of P_1, the trajectories of (5) tend to this point,
   - is a saddle point if $\mu > \gamma/k_1\sigma$ whose two dimensional stable manifold is in the ah plane. Therefore, for small enough population densities, plants and herbivores become extict and the pollinators population stabilizes at K.

3. for an appropriate parameter values, the point $\bar{P}$ is locally asymptotically stable.

Following [7], let us introduce the following definition.

Definition. *The system $\dot{x} = F(x)$ is called* persistent, *if* $\lim_{t\to\infty} x_i(t) > 0$ *where* $x(t) = (x_i(t))_{1\leq i\leq n}$ *is any solution with* $x_i(0) > 0$ *for* $1 \leq i \leq n$. *The system is* uniformly persistent, *if there exists* $d > 0$ *such that* $\lim_{t\to\infty} x_i(t) \geq d$ *for* $1 \leq i \leq n$ *and for any solution with positive initial conditions.*

In [7], the author proves the following theorem.

Theorem 3.

1. *The solutions of the system (5) starting from (a_0, p_0, h_0) with a_0, p_0 and h_0 greater or equal zero, are not negative and bounded for all t,*
2. *Let $\mu > \mu_1$, then the system (5) is uniformly persist if $h(\bar{p}^*) > \delta/m_2$,*
3. *All solution of the system with $a(0) > 0$ and $p(0) > 0$ converges to $(\bar{a}^*, \bar{p}^*, 0)$ if $h(\bar{p}^*) < \delta/m_2$.*

Case 2. $g(h) = \frac{1}{1+k_3 h^2}$ with $k_3 > 0$. Here we present a simple reasoning which shows us that, for appropriate parameter values, the system (5) has just an equilibrium, (a^*, p^*, h^*), in the positive octant of the *aph* space. For this aim, let us simplify the notation in (5) by introducing β, α_1 and α_2 as follows

$$\beta = \sigma\phi\mu^2, \quad \alpha_1 = k_2\sigma\mu \text{ and } \alpha_2 = k_1\sigma\mu.$$

From the third equality of (6) we obtain $p = p^* = c_1\delta/(m_2 - \delta)$ which, from an interpretative point of view, it is necessary to impose the condition $m_2 > \delta$. For each value, p^*, of p, the equality $p = p^*$ defines a plane parallel to the plane ah in the space phase aph of the system (5). In these terms, searching the equilibrium points of (5), is "reduced" to look at the behaviour of the projections of the null-clines on the plane $p \equiv p^*$. Thus, for fixed p^*, we have that the equality $f_1 = 0$, implies the following condition on h^* and a^*

$$g(h^*) = \frac{(a^* - K)(1 + \beta p^*)}{\alpha_1 p^*},$$

where $a^* > K$. For the particular form $g(h) = 1/(1 + k_3 h^2)$, as it is required, from the previous equality we find a first relationship between h^* and a^*. This is

$$h_1^*(a^*) = \sqrt{k_3 \frac{\alpha_1 p^* - (a^* - K)(1 + \beta p^*)}{(a^* - K)(1 + \beta p^*)}}, \tag{7}$$

which, defines positive values of h, whenever in addition to $a^* > K$, the condition

$$K < a \leq K + \frac{\alpha_1 p^*}{1 + \beta p^*}.$$

holds. From the equality $f_2 = 0$, we obtain a second relationship between h^* and a^*. This is

$$h_2^*(a^*) = \frac{c_1 + p^*}{m_1 p^*} \left[\frac{\alpha_2}{\alpha_1} a^*(a^* - K) - \gamma \right]. \tag{8}$$

Therefore, for fixed p^*, the amounts a^* and h^* must satisfy (7) and (8). Figure 11 shows us the behaviour of h_1^* and h_2^* as functions of a^* for different values of the parameter K. From these, one can see that —for the appropriate other parameter values— for each K, such graphs touch each other in the first positive quadrant in just one point, implying

127

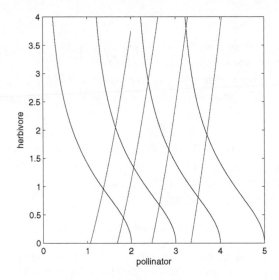

FIGURE 11. Graphs of $h_1^*(a^*)$ (in black) and $h_2^*(a^*)$ (in blue) corresponding to increasing values of the parameter K. Those are the projections on the plane $p = p^*$. See the text for details.

that the system (5) has exactly one equilibrium in the first positive octant, as we claimed previously.

On the basis of the existence of a positive equilibrium for the system (5), we carried out a set of numerical simulations in order to obtain the space phase[4] of it corresponding to different values of the parameter K. What we noted is the existence of a rich temporal dynamics which includes: the existence of a homoclinic trajectory based at the equilibrium $P = (K, 0, 0)$, breaking of a homoclinic loop, emergence of limit cycles, the existence of a global attractor. These are shown in figures 12 and 13.

The spatio-temporal dynamics

Here we consider the following hypothesis: 1. The pollinators movement has two component: its "own" at random individual movement and a drift (advection) due to, for instance, the wind. In a first approximation, we are going to consider a constant advection velocity, $\vec{v}$, 2. At individual level, the herbivores move at random; their movement is much slower than that of the pollinators. This is reasonable when one thinks in terms bees as pollinators and cows as herbivores, for instance, 3. The spatial distribution of the plants changes because of the interaction with the pollinator and herbivore populations, 4. The temporal dynamics of the interaction, is described by the system (5). One model

---

[4] All the three dimensional space phase we present here, were done by a using *MATLAB 7.0*.

128

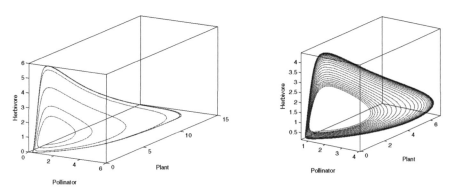

FIGURE 12. Phase portrait of the system (5) in the Case 2, corresponding to $K = 0.25$ (left) and $K = 0.65$ (right).

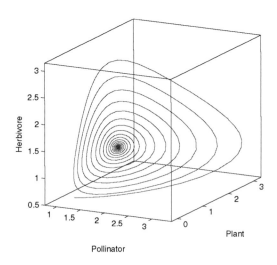

FIGURE 13. For $K = 1$

reflecting these assumptions is:

$$
\begin{aligned}
\frac{\partial a}{\partial t} &= D_1 \nabla^2 a - \vec{v} \cdot \nabla a + ba(K - a) + \frac{g(h)k_2 \sigma \mu ap}{1 + \sigma \phi \mu^2 p} \\
\frac{\partial p}{\partial t} &= -\gamma p + \frac{g(h)k_1 \sigma \mu ap}{1 + \sigma \phi \mu^2 p} - \frac{m_1 ph}{c_1 + p} \\
\frac{\partial h}{\partial t} &= D_2 \nabla^2 h - \delta h + \frac{m_2 ph}{c_1 + p},
\end{aligned}
\tag{9}
$$

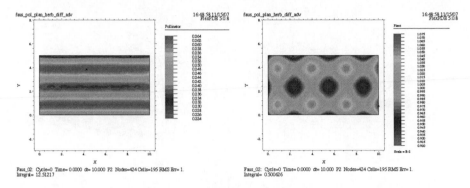

FIGURE 14. Numerical solutions of an initial and boundary values problem associated with the rda system (9). Initial conditions for pollinator (left) and plant (right).

where $g(h) = 1/(1+k_3 h^2)$, plus the initial (on Ω) and homogeneous Neumann boundary (on $\partial\Omega$) conditions, completes the mathematical problem to being analized.

We carried out two sets of numerical simulations in both of them we used a rectangular habitat with

$$\vec{v} = (v,0) \text{ with } v = 3, \ D_1 = 3, \ D_2 = D_1/10.$$

Numerical simulations 1. Here we consider the parameter values $(K = 0.25)$ for which our space phase simulations suggest the existence of a homoclinic trajectory based at $(K,0,0)$. The initial conditions are

$$a(\vec{r},0) = 0.25(1+0.1\sin(2y)), \ \ p(\vec{r},0) = 0.01(1+0.1\sin(x+y)\cos(x-y)),$$

and

$$h(\vec{r},0) = 0.01(1+0.1\sin(2x)),$$

which corresponds to a perturbation of the steady and homogeneous state associated with the equilibrium $(K,0,0)$. Figures 14-20 shows our results.

From them we can see that the above mentioned initial conditions subject to the spatio-temporal dynamics given by the system (9) —after some transients— evolve towards a sort of shifted band pattern which, for the pollinators and herbivore, travels from left to the right; meanwhile for the plants, visually it corresponds to a similar pattern but, given that those do not move then, the plants once than are consumed by the herbivores, they regenerate and do so in such a way that the visual effect is a moving pattern. See the figure 7.

Numerical simulations 2. Here we choosed the parameter for which our numerical simulations suggest us that the positive equilibrium is an attractor for the trajectories of system (5) in the first octant. By taking as initial conditions a perturbation of the

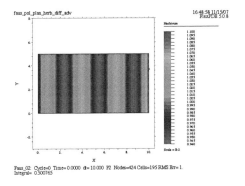

FIGURE 15. Initial conditions for herbivores.

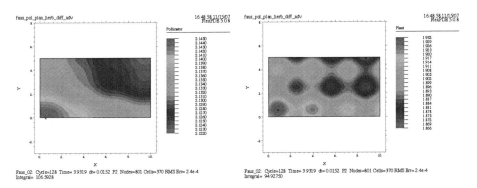

FIGURE 16. A transient spatial distribution for pollinator (left) and plant (right).

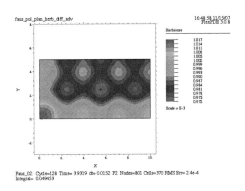

FIGURE 17. For herbivores.

131

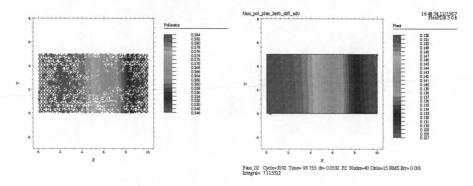

FIGURE 18. Final distribution for pollinator (left) and plant (right).

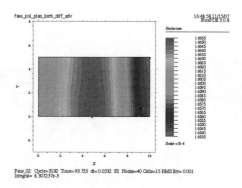

FIGURE 19. For herbivores.

steady and homogeneous state corresponding to that equilibrium, the asymptotic pattern is the homogeneous spatial distributions for the three populations given by the respective coordinates of the positive equilibrium.

CONCLUSIONS AND DISCUSSION

We must say that much of the material we presented in this paper, particularly that dealing with the temporal and spatio-temporal dynamics for the pollinator-plant-herbivore interaction, still under current investigation. Here we presented just some preliminary results. In this sense, we should say that at the present stage of our project, we have more questions than answers. We organize the conclusions of this paper and the dicussion in the following items:

1. The models we studied in this paper, are not based on any specific field ecological data. However, all of the interacting terms —particularly those of Holling response type– are abundantly refered in the ecological literature; meanwhile those trying

to reflect the strictly spatial dynamics (diffusion and advection) were proposed on the basis of reasonable assumptions. In spite of this lack, the analysis we did gives us interesting insights of the type of spatio-temporal dynamics these systems can support.

2. **Functional response of type IV.** The Holling response of type II was abundantly used through this paper. However, it is documented in the ecological literature (see [2] and [4]) that for "small" enough values of prey density, the predation rate per unit of predator, is a monotone increasing function of the prey density, but for high prey density, it becomes a monotone decreasing function. This is due to a sort of prey defense strategy which only opers at high population densities. In this case, rather that having a monotonic increasing function having an asymptotically behaviour as that used along this paper, we should have a nonmonotonic Hollins response which could have one of the following analytical forms:

$$\varphi(x) = \frac{x}{1+x}e^{-kx}, \ \text{or} \ \varphi(x) = \frac{x}{a+bx+cx^2},$$

where k, a, b and c are positive parameters. Thus, we consider that it is worth to investigate the influence of this functional response in the spatio-temporal dynamics of interacting species models.

The temporal and the spatio-temporal dynamics involved in this case is currently studied by one of us [13] and [21].

3. **Density-dependent diffusion terms or plant-taxis.** Here the idea is to consider a more realistic situation in which the pollinator and herbivore movement is towards those places of the habitat where there are plants. In such a case, instead of having constant diffusion coefficients for the the pollinator and herbivore populations as we did it here, these should be plant population dependent.

4. We consider that it is worth to investigate the conditions for the initial and boundary value problem associated with the system (9), for which such a system supports spatial patterns and classify them.

REFERENCES

1. D.H. Boucher (1982): *The Biology of Mutualism*. Oxford University Press.
2. J.B. Collings (1997): The effect of the functional response on the bifurcation behaviour of a mite predator-prey interaction model. J. Math. Biol., **36**, pp. 149-168.
3. I.D. Couzin and J. Krause: *Self-organization and collective behaviour in vertebrates*. Advances in the Study of Behavoiur. In press.
4. M.J. Crawley (1992): *Natural enemies: The population biology of predators, parasites, and disease*. Blackwell Scientific Publications., Oxford.
5. G. García-Ramos, F. Sánchez-Garduño and P.K. Maini (2000): Dispersal can sharpen parapatric boundaries in a spatially varying environment. Ecology, **81**, No. 3, pp. 749-760.
6. E. E. Holmes, M. A. Lewis, J. E. Banks, and R. R. Veit (1994): Partial Differential Equations in Ecology: Spatial Interactions and Population. Ecology, **75**(1), pp. 17-29.
7. S.R. Jang (2002): Dynamics of herbivore-plants-pollinator models. J. Math. Biol., **44**, pp. 129-149.
8. M. Kot (2001): *Elements of mathematical ecology*. Cambridge University Press.
9. Y. A. Kuznetsov (2004): *Elements of Applied Bifurcation Theory*. Applied Mathematical Series 112. Springer-Verlag.

10. A. Okubo (1980): *Diffusion and Ecological Problems: Mathematical models*. Berlin Heidelberg New York. Springer-Verlag.
11. A. Okubo (1986): Dynamical aspects of animal grouping: Swarms, schools, and herds. Adv. Biophys, **22**, pp. 1-94.
12. A. Okubo and S.A. Levin *Diffusion and Ecological Problems, modern perspective*. Springer 2001.
13. I. Quilantán (2007): Dinámica espacio-temporal de una interacción polinizador-planta-herbívoro. Tesis de Maestría en Matemáticas Aplicadas (director: Faustino Sánchez Garduño), DACB, UJAT. Work in progress.
14. F. Sánchez-Garduño (2001): Continuous density-dependent diffusion modelling in ecology: A review. Recent Res. Ecol., **1** pp. 115-127.
15. F. Sánchez-Garduño, P.K. Maini and J. Pérez-Velázquez: A nonlinear degenerate equation for direct aggregation. In preparation.
16. J.G. Skellam (1951): Random dispersal in theoretical populations. Biometrika, **38**, pp. 196-218.
17. J.G. Skellam (1973): The formulation and interpretation of mathematical models of diffusionary processes in population biology. In: *The mathematical theory of the dynamics in biological populations*. M.S. Batchellet *et al* Editors. Academic Press.
18. J. Soberón M. and C. Martínez del Río (1981): The dynamics of a plant-pollinator interaction. J. Theor. Biol., **91**, pp. 363-378.
19. P. Turchin and P. Kareiva (1989): Aggregation in *aphis varians*: an effective strategy for reducing risk. Ecology **70**(4), pp. 1008-1016.
20. P. Turchin (1998): *Quantitative Analysis of Movement: Population redistribution in animals and plants*. Sinauer, Sunderland. M.A.
21. G. Velázquez (2007): Dinámica temporal de una interacción polinizador-planta-herbívoro. Tesis de Maestría en Matemáticas Aplicadas (director: Faustino Sánchez Garduño), DACB, UJAT. Work in progress.

A Biophysical Neural Model To Describe Spatial Visual Attention

Etienne Hugues[1] and Jorge V. José[1,2]

(1) Department of Physics, (2) Department of Physiology and Biophysics,
SUNY at Buffalo, Buffalo, NY 14260.

Abstract. Visual scenes have enormous spatial and temporal information that are transduced into neural spike trains. Psychophysical experiments indicate that only a small portion of a spatial image is consciously accessible. Electrophysiological experiments in behaving monkeys have revealed a number of modulations of the neural activity in special visual area known as V4, when the animal is paying attention directly towards a particular stimulus location. The nature of the attentional input to V4, however, remains unknown as well as to the mechanisms responsible for these modulations. We use a biophysical neural network model of V4 to address these issues. We first constrain our model to reproduce the experimental results obtained for different external stimulus configurations and without paying attention. To reproduce the known neuronal response variability, we found that the neurons should receive about equal, or balanced, levels of excitatory and inhibitory inputs and whose levels are high as they are in *in vivo* conditions. Next we consider attentional inputs that can induce and reproduce the observed spiking modulations. We also elucidate the role played by the neural network to generate these modulations.

Keywords: Neural networks, Vision, Attention.
PACS: 87.19.L, 87.19.ll, 87.19.lc,

INTRODUCTION

The neural basis of many cognitive brain capabilities has been intensively studied in recent years using a variety of experimental techniques. Among these, recording the electrical activity of individual neurons in awake animals while performing a prescribed task has given insight on the way neurons respond inside a given brain when performing a given task. However, despite these invaluable results, it has not been possible to understand the underlying neuronal mechanisms leading to this type of response, because a single neuron integrates synaptic inputs from thousands of other neurons. Typically a neuron interacts with its postsynaptic neurons when it emits an action potential -a voltage spike of its membrane potential, which propagates along its axon and activates its synapses and releasing neurotransmitters that modify the activity of the postsynaptic neurons. Although these spikes are thought to be the mechanism by means of which information is exchanged between neurons, it remains difficult to infer the specific way information about the task is encoded in this electrical activity.

Accurate empirical neuronal and synaptic models have been developed that open the way as to how to describe this electrical activity, going from single neurons to

CP978, *Biological Physics, 3rd Mexican Meeting on Mathematical and Experimental Physics*
edited by L. Dagdug and L. García-Colín Scherer
© 2008 American Institute of Physics 978-0-7354-0497-7/08/$23.00

networks of neurons representing specific brain areas. Solving these highly non-linear models numerically one may be able to reproduce available experimental data. Once the modeling results agree with the experimental data it may be possible to infer the underlying mechanisms at play, and the change in behavior while varying the diverse parameters and deduce the way neural information is encoded. Building a theoretical understanding can lead to novel predictions which can be tested experimentally.

In this paper we will focus on the problem of *selective visual attention*, and particularly in spatial visual attention, i.e. the ability to pay attention towards a precise location in the visual field. The experiments we are interested in are being done with behaving macaque monkeys and they measure the V4 neural activity in their visual system, where the attentional modulations have been found to be important [1].

The standard method to measure the electrical spike activity of neurons in awake monkeys is to use extracellular electrodes. The timing and the number of these spikes have been found to be variable from trial to trial for a given neuron. The neural response is typically quantified by its mean firing rate, i.e. the mean number of spikes per unit time across trials. Appropriate neurons are chosen for study such that their receptive fields (RF) locations are found, i.e. the region of the visual field to which the neuron spikes, to a desirable location. The stimuli presented have a given orientation (like oriented bars), and the neurons respond selectively to this orientation: if all possible orientations of the same stimulus are presented, the resulting firing rate bell shape curve, called the tuning curve (TC), is determined with its maximum indicating the preferred neural orientation.

A typical task is shown in Fig. 1. The monkey is trained to fixate at a central point on a screen, then needs to pay attention to a given location indicated by a cue [1-6] or to given moving stimuli [7]. The same conditions are repeated a number of times to obtain meaningful statistical results. In some of the trials, when a stimulus is presented inside the RF, attention can be directed inside the RF ("attention in") or outside ("attention out"). It was found that generally the mean firing rate increases when attention is directed inside the RF [2-3,5,7]. Furthermore, the TC of a neuron was found to be modulated by a multiplicative factor [3].

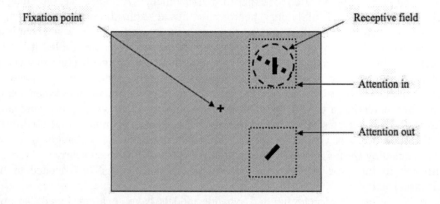

FIGURE 1. Typical visual experimental task.

Neurons in V4 have a sufficiently large RF that two stimuli can be presented at the same time. This property raises the question as to how a neuron does respond to the superposition of two objects and how attending to one of them modifies the response? It has been found that the neuron's firing rate response when both stimuli are present is in between the response to each stimulus when each is presented alone [4]. This finding is called *stimulus competition* [2]. When one stimulus is attended, the neuron's response was biased [2] and moved toward the response elicited by this stimulus alone [4].

Extracellular recordings also measure the local field potential (LFP). This is a signal which represents the local electrical activity in the neural tissue around the electrode. The frequency signal is in the range of a few to hundreds of Hz, which indicates the presence of these rhythms in the local neural population. Attention has been found to enhance the LFP oscillation in the gamma frequency range (30-80 Hz) and also to enhance the spike lockings with this same type of oscillation [5-6].

In a recent paper in an *in vivo* experiment it has been possible to distinguish between different neural classes just on the basis of their extracellular spike waveforms [7]. Also, based on clear evidence obtained in *in vitro* recordings, broad spike waves can be attributed to pyramidal excitatory neurons (PN), and narrow spikes attributed to parvalbumin responsive inhibitory interneurons (IN). These results have shown that INs fire much faster than PNs. The response neural variability, measured by the Fano factor FF (defined by the ratio of the variance to the mean of the spike count for a given time interval) shown to be large ($FF \approx 1.4$). Of importance is that these results provide a precise characterization of the neural activity with or without stimulation and/or attention across different neuronal populations.

These experiments have shown that attention induces different types of neural activity modulations. As pointed above, it is not known from these results why these different modulations occur and what role do they play. It is also not known if these modulations are related to each other and how, or if they can be explained in a unique theoretical framework. We believe that, for a model to be able to capture all these phenomena, a neural network description of V4 is needed [8]. Our hypothesis is that this network is sufficient, provided it receives the inputs equivalent to those received *in vivo*. Taking into account the experimental observations, we will show here how our model can reproduce many of these results in a unified way. Some of the results presented here have been previously published in a Society of Neuroscience abstract in 2007 [9].

THE MODEL

The experiments involve the spatial and the orientation neural selectivities, which means that at least a two-dimensional neural layer is necessary to reproduce the experimental results. Assuming that the connectivity between neurons selective to different locations is similar to the connectivity between neurons selective to different orientations, we can then infer, via a proper normalization in the two directions, that the distance between these neurons within the layer is the only parameter that matters. For convenience, and computing simplification, we only model a one-dimensional approximation of this layer. It is still possible to consider the presentation of two stimuli at different locations and with different orientations in the simplified model. We assume that this approximation will lead to correct qualitative results but not accurate quantitatively. The one-dimensional layer of neurons will be assumed to have a ring topology. In this ring, neurons are spatially uniformly distributed, with a neuronal x value between 0 and 1 assigned to each one of them, characterizing the preferred neuronal stimulus. If we considered only superposed stimuli, the ring would be biologically justified, and x would represent the preferred orientation of each neuron.

The Network

The network, shown in Fig. 2, is composed of N_E pyramidal neurons (PN) and N_I inhibitory neurons (IN), with $N_E = 4N_I$, similar to the ratio proportions found in the brain cortex. In the results presented here the simulations were done with $N_E = 400$ and $N_I = 100$.

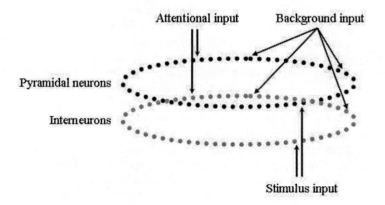

FIGURE 2. Sketch of the network model.

To help reproduce the experimental results quantitatively, we consider biophysical models to describe neurons and synapses. PNs are represented by the Ermentrout-Kopell model [10] and INs by the Wang-Buzsaki model [11]. Both models are mathematically simplified versions of the Hudgkin-Huxley conductance-based models. In both type of models, the membrane potential V obeys to the following differential equation:

$$ C\frac{dV}{dt} = -g_L(V - V_L) - g_{Na}m^3h(V - V_{Na}) - g_K n^4(V - V_K) + I_{syn}. $$

138

C is the membrane capacitance. The three first terms on the right are the membrane currents: the leak (L), the sodium (Na) and the potassium (K) currents. The synaptic current, I_{syn}, is the sum of all currents received at the neuron's synapses, which are either input synapses or network synapses. For the membrane current X ($X = L,Na,K$), g_X is the maximal conductance and V_{syn} is the reverse potential. m and h are, respectively, the activation and inactivation sodium current variables, and n is the activation variable of the potassium current. In the PN model, $m = a_m(V)/[a_m(V) + b_m(V)]$ and $h = \max(1 - 1.25n, 0)$, with n obeying to the following kinetic equation:

$$\frac{dn}{dt} = a_n(V)(1-n) - b_n(V)n.$$

In the IN model, $m = \alpha_m(V)/[\alpha_m(V) + \beta_m(V)]$ and h and n obey to the equations:

$$\frac{dh}{dt} = \phi[\alpha_h(V)(1-h) - \beta_h(V)h]$$

$$\frac{dn}{dt} = \phi[\alpha_n(V)(1-n) - \beta_n(V)n].$$

Further details about these models can be found in the Appendix.

The synaptic current entering a postsynaptic neuron via a synaptic receptor is given by

$$I_{syn} = -g_{syn}s(t)(V - V_{syn}),$$

where g_{syn} is the maximal synaptic conductance; $s(t)$ is a gating variable representing the proportion of opened synaptic channels at time t; V is the potential of the postsynaptic neuron and V_{syn} is the reverse potential of the synapse ($V_{syn} = 0\,mV$ for excitatory synapses and $V_{syn} = -70\,mV$ for inhibitory synapses). For AMPA and GABA$_A$ receptors, the gating variable is given by

$$s(t) = \sum_k H(t - t_k)(e^{-(t-t_k)/\tau_d} - e^{-(t-t_k)/\tau_r})$$

where the sum is over all incoming spikes. H is the Heaviside step function ($H(t) = 1$ when $t \geq 0$, otherwise $H(t) = 0$). For an AMPA receptor, the decay time is $\tau_d = 2\,ms$ [12-13] and the rise time is $\tau_r = 1\,ms$. $\tau_r = 1ms$. For a GABA$_A$ receptor, $\tau_d = 10\,ms$ [14-15] and $\tau_r = 1\,ms$. For an NMDA receptor, the current has the voltage dependence $1/\{1 + [Mg^{2+}]e^{-0.062V}/3.57)\}$, which is controlled by the external magnesium concentration $[Mg^{2+}]$ [16] (here, $[Mg^{2+}] = 1\,mM$). The gating variable is obtained from solving the following set of equations:

$$\frac{ds}{dt} = -\frac{s}{\tau_d} + \alpha x(1-s)$$

$$\frac{dx}{dt} = -\frac{x}{\tau_r} + \sum_k \delta(t-t_k),$$

where δ is the Dirac-delta function, $\tau_d = 100\,ms$, $\alpha = 0.5\,ms^{-1}$ and $\tau_r = 2\,ms$.

The neurons are assumed to be connected at random. A neuron at position x in the population X is connected to a neuron at position y in the population Y with probability

$$p_{XY}(x,y) = \frac{P_{XY}}{\sqrt{2\pi}\sigma_{XY}} e^{-d^2/2\sigma_{XY}^2}.$$

Here d is the minimum distance between the neurons on the ring. The width of the connectivity profile σ_{XY} is chosen to be small compared to 1 ($\sigma_{XY} \le 0.15$), so that P_{XY} is almost the mean connection probability between populations X and Y. The connectivity is chosen to be sparse ($P_{XY} \le 0.1$).

The Inputs

V4 receives different inputs from different parts of the brain: the visual stimulus, the attentional stimulus, and also inputs from other brain areas connected to V4 and which are active even if this activity is unrelated to the attention task.

The visual stimulus is predominantly coming from area V2, and we will consider that this is the only stimulus source to V4. This input is localized around x_s, with a Gaussian shape and width σ_s. Even if the top-down attentional input is thought to originate in the prefrontal cortex or in the parietal cortex, the precise area -if there is only one- is not known for sure. However, we don't need to know where this signal originates in our model, but we need to know what is its essential nature, which is also mostly unknown. This input is mediated by feedback connections, which are known to be essentially excitatory [17-18]. Experiments have suggested that this input is local [19] and that it may oscillate in the gamma frequency range [20]. We name A the area from which the input originates. When attention is paid at location x_a, we consider that the input change from A is localized around x_a and oscillates at the gamma frequency f_a (here, $f_a = 50\,Hz$ [5-6]). We will consider the case when the mean excitation does not change and also the case when it increases. All areas, including V2 and A, produce a background input to V4 which, in absence of a stimulus or attention, induces a spontaneous V4 neural activity at low firing rates.

All external inputs to the network are modeled as Poisson spike trains, characterized only by their frequency. In what follows, we will describe our search for the quantitative values of these inputs so as to reproduce the experimental data.

RESULTS

The model involves a number of synaptic and input parameters whose values are not known. Using the experimental data obtained for various stimulus presentations in the unattended condition, we have been able to choose realistic biophysical parameter values for which the model reproduces the data. Next we present the results or our attentional studies under different attentional inputs on network activities.

Network Response To Stimulus Presentations

One stimulus case

Either spontaneously or in response to the presentation of a stimulus, INs fire more rapidly than PNs [7], but the neural response variability, measured by the Fano factor, is similarly high for both neuron types. A number of different inputs can reproduce the observed firing rates. However, the response variability is known to be high only when the excitatory and the inhibitory components of a neuron's input are close to be balanced, which is thought to constrain the neuron to fire at a low frequency. To find the right region in parameter space, we have studied the firing response of a single PN and a single IN as a function of its excitatory and inhibitory inputs. We considered input spike trains with Poissonian statistics. This is a very good approximation since external inputs in the model also have this statistics. The recurrent input -originating from the network- is also found to have a statistics very close to Poisson (data not shown). In this case, the mean excitatory and inhibitory currents are proportional to their frequency inputs. We found that the FF is high ($FF > 1$) in the region when the input is balanced, i.e. when the ratio of excitatory to inhibitory input frequencies stays approximately constant. More interestingly, we found that the region of high FF widens as input frequencies increase, and that the maximum firing rate increases too. We conclude that the elevated firing rates as seen during stimulus presentations together with high FF are only possible when input rates are high and at levels comparable to *in vivo* values.

To test this single neuron prediction, we have simulated the network spontaneous response (data not shown) and its response to one stimulus (see Fig. 3) when neurons receive high frequency total inputs. The network activity reproduces the firing rates seen in vivo, with a bump of elevated activity for PNs and INs around the stimulus location x_s. As suggested by the single neuron analysis, both network populations of neurons have a high response variability.

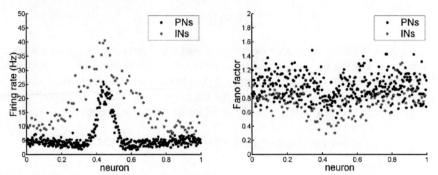

FIGURE 3. One stimulus presentation network response at x_s=0.45. *Left*: Firing rates for PNs (*black dots*) and INs (*grey dots*). *Right* : Fano factor for PNs and INs.

Two stimuli case

When two stimuli are presented simultaneously, the response of neurons in V2 has been found to be between the response obtained when both stimuli are presented separately [4]: therefore, there is also stimulus competition in V2. In our model it means that the V2 input to V4 must be less than the input sum from each stimulus. This lower input level is shown to lower the response in V4, as compared to the case where there is no such competition in V2. The spatial extent of inhibitory connections in the network, which leads to a lateral inhibition, contributes to lower the response in V4, because more inhibitory neurons are activated when two stimuli are presented. We conclude that stimulus competition in V2 and lateral inhibition in V4 lead to stimulus competition in V4.

The results of the neural network response to the simultaneous presentation of two stimuli are shown in Fig. 4. The neurons that respond to any stimulus presentation have an intermediate firing rate when both stimuli are presented. For a given neuron, the stimulus presentation S_n ($n=1,2$) leads to a firing rate response f_n, and the presentation of stimuli S_1 and S_2 leads to a firing rate response f_{12}. If we plot the sensory interaction $SI = f_{12} - f_1$ as a function of the selectivity $SE = f_2 - f_1$, we observe that (see Fig. 4) for all the responding neurons, the points tend to be aligned. As observed in experiment [4], the best line-fit (in the least-square sense) has a slope less than one and intersects the SI-axis at a small value.

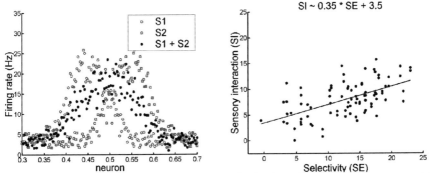

FIGURE 4. *Left*: Firing rates of PNs in response to each stimulus and to the simultaneous presentation of both stimuli. *Right*: sensory interaction as a function of selectivity for all responding PNs. The best line-fit equation is given above the graph.

Attentional Modulations Of Network Activity

One stimulus case

We first consider the case for which attention does not change the mean level of excitation originating from A: in the attended condition, the input from A oscillates at frequency f_a around the location x_a, with its temporal mean the same as in the unattended condition. The result of the neural network response when a stimulus S presented at x_s is attended ($x_a = x_s$) is shown in Fig. 5. For sufficient but relatively weak oscillation amplitude, the attentional input induces a firing rate increase for the PN population around x_s, which decreases as the difference between a neuron's preferred stimulus x and x_s increases. Because the inputs to the neurons are almost balanced, adding an oscillatory input of sufficient amplitude increases their firing probability when the balance is changed temporarily in favor to excitation.

This increase in the neuron's response when attention is directed inside the neuron's RF is what has been found experimentally [2-3,5]. This modulation is responsible for the gain modulation of the TC of a neuron (data not shown) [3].

If we consider the case for which attention increases locally the mean level of excitation originating from A, we find that the firing rate modulation can be due only to this mean excitation increase (data not shown). In this case, the input oscillation only contributes to the network oscillation.

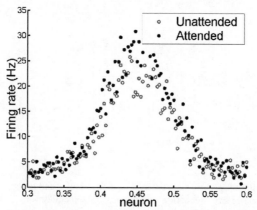

FIGURE 5. PN population firing rate when a stimulus at $x_s=0.45$ is not attended (*grey circles*) or when it is attended (*black dots*). The firing rate modulation appears around the attended location $x_a=x_s$.

Two stimuli case

When two stimuli S_1 and S_2 are simultaneously presented at x_1 and x_2 in the network, and when stimulus S_n is attended ($x_a = x_n$), there is a PN population firing rate increase, just around x_a, but also a further decrease. The increase is again due to the attentional signal, as in the one stimulus case, but the decrease is due to the lateral inhibition in the network. As observed in the experiments [4], the slope of the best line-fit increases when the stronger stimulus is attended, and it decreases when the weaker stimulus is attended (see Fig. 6).

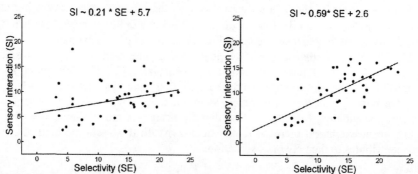

FIGURE 6. Sensory interaction as a function of selectivity for all responding PNs. The best line-fit equation is given above the graph. *Left*: PNS for which the weaker stimulus is attended. *Right*: PNs for which the stronger stimulus is attended.

In the unattended condition, we have chosen an asynchronous network state as indicated by the experimental results where the gamma frequency range amplitude oscillation is weak [5-6]. Our approximation here is thus reasonably accurate. In the attended condition, the oscillatory attentional input induces an oscillation at the same frequency, locally around the attended location x_a. In this region, neurons still have irregular spike trains, but the distribution of their firing phases with respect to the input oscillation is no longer flat. This is illustrated in Fig. 7 where we show the Fourier transform power of these spike trains (which grows from 0 to 1 going from completely asynchronous to completely locked spike train state) which is weak but also shows a marked peak at the input frequency. This result is analogous to the increase of the spike-field coherence observed experimentally [5-6].

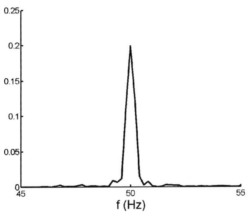

FIGURE 7. Fourier transform neural spike train power for neurons whose stimulus preference x is close to the attended location x_a. The peak shows that the network oscillates locally at the input frequency.

DISCUSSION

In this paper we have presented results of a neural network model calculations of different mechanisms that attempt to explain the important problem of paying attention. We were motivated by recent experimental results carried out both in primates and in *in vitro* slice experiments that have found important neuronal signatures of this cognitive effect. We have introduced a neural network model to describe the visual area V4. We calibrated the model parameters so as to reproduce the experimental results obtained for different stimulus configurations in the unattended condition. We found that, for all neurons to exhibit high response variability, their excitatory and inhibitory inputs must approximately be balanced as it is found in *in vivo* conditions. We find that stimulus competition occurs in the V4 network both because stimulus competition exists in V2 and because of lateral inhibition in the network. We found that different forms of attentional inputs can induce the experimentally observed modulations: oscillate in the gamma frequency range with a mean excitation which can be either unchanged, or increase in the attended condition. In both cases, the relatively weak attentional input is able to increase the local neuron firing rate because these neurons receive an approximately balanced input, and the input oscillation induces a local neural oscillation in the network. When two stimuli are presented and one is attended, the network response to this stimulus increases and lateral inhibition induces a decrease in the network response for the unattended stimulus, leading to a bias competition.

In conclusion, our neural network model is able to reproduce a range of experimental observations in one unified way. In future studies we will study in more detail the effects of changing the parameters of the attentional signal. We will also try to reproduce other experimental attentional results.

The one dimensional network model proposed here is a simplification of the real neuronal circuits in V4. The physiological circuits are organized in several layers and contain more than two neuron types. However, this simplified type of model has proven to be very useful in shading light in the understanding of basic neuronal mechanisms present in the visual brain areas, such as the primary visual area V1.

A number of other network models have been recently proposed to address the attentional neural mechanisms [21-24]. Many of these models lack, however, sufficient physiological realism in structure or in the neuronal network [21-23]. A recent neural network study has addressed a different type of attention in visual area MT [24]. Although some of the experimental results obtained in MT have similarities with those found in V4, there are important differences. That study has made the hypothesis that a prefrontal or parietal area is reciprocally connected with MT and that is the attentional input source, therefore constraining the form of the attentional input to MT. In the V4 case, the source of the attentional signal is still debated, thus we concentrated on possible different forms of the attentional input regardless of whatever its source may be.

The results we are presenting here are preliminary and a more detailed description will be published elsewhere.

APPENDIX

The following units are used: a capacitance is expressed in $\mu F / cm^2$, a conductance in mS / cm^2 and a potential in mV.

In the PN model [10], we have taken $C = 1$, $g_L = 0.1$, $V_L = -67$, $g_{Na} = 100$, $V_{Na} = 50$, $g_K = 80$ and $V_K = -100$. The functions that define the activation variables are given by,

$$a_m(V) = 0.32 \frac{V + 54}{1 - e^{-(V+54)/4}}$$

$$b_m(V) = 0.28 \frac{V + 27}{e^{(V+27)/5} - 1}$$

$$a_n(V) = 0.032 \frac{V + 52}{1 - e^{-(V+52)/5}}$$

$$b_n(V) = 0.5 e^{-(V+57)/40}.$$

In the IN model [11], we took $C = 1$, $g_L = 0.1$ and $V_L = -65$, $g_{Na} = 35$, $V_{Na} = 55$, $g_K = 9$, $V_K = -90$ and $\phi = 5$. The corresponding functions defining the activation/inactivation variables are:

$$\alpha_m(V) = 0.1 \frac{V + 35}{1 - e^{-(V+35)/10}}$$

$$\beta_m(V) = 4 e^{-(V+60)/18}$$

$$\alpha_h(V) = 0.07 e^{-(V+58)/20}$$

$$\beta_h(V) = \frac{1}{1 + e^{-(V+28)/10}}$$

$$\alpha_n(V) = 0.01 \frac{V + 34}{1 - e^{-(V+34)/10}}$$

$$\beta_n(V) = 0.125 e^{-(V+44)/80}.$$

ACKNOWLEDGMENTS

We thank UB funding support for the work presented here as well as conversations with Paul Tiesinga and Scott Hill.

REFERENCES

1. Moran, J., and Desimone, R., *Science* **229**, 782-784 (1985).
2. Desimone, R., and Duncan, J., *Annu. Rev. Neurosci.* **18**, 193-222 (1995).
3. McAdams, C., and Maunsell, J., *J. Neurosci.* **19**, 431-441 (1999).
4. Reynolds, J. H., Chelazzi, L., and Desimone, R., *J. Neurosci.* **19**, 1736-1753 (1999).
5. Fries, P., Reynolds, J. H., Rorie, A. E., and Desimone, R., *Science* **291**, 1560-1563 (2001).
6. Womelsdorf, T., Fries, P., Mitra, P. P., and Desimone, R., *Nature* **439**, 733-736 (2006).
7. Mitchell, J. F., Sundberg, K. A., and Reynolds, J. H., *Neuron* **55**, 131-141 (2007).
8. Hugues, E., and José, J. V., *Int. J. Mod. Phys. E*, in press.
9. Hugues, E., and José, J. V.. "Spatial Attention Mechanisms In V4: A Biophysical Network Model", *Soc. Neurosci. Abs. 636.12*, 2007.
10. Ermentrout, G. B., and Kopell, N., *Proc. Nat. Acad. Sci.* **95**, 1259-1264 (1998).
11. Wang, X. J., and Buzsaki, G., *J. Neurosci.* **16**, 6402-6413 (1996).
12. Hestrin, S., Sah, P., and Nicoll, R., *Neuron* **5**, 247-253 (1990).
13. Spruston, N., Jonas, P., and Sakmann, B., *J. Physiol.* **482**, 325-352 (1995).
14. Salin, P. A., and Prince, D. A., *J. Neurophysiol.* **75**, 1573-1588 (1996).

15. Xiang, Z., Huguenard, J. R., and Prince, D. A., *J. Physiol.* **506**, 715-730 (1998).
16. Jahr, C.E., and Stevens, C. F., *J. Neurosci.* **10**, 3178-3182 (1990).
17. Johnson, R. R., and Burkhalter, A., *J. Comp. Neurol.* **368**, 383-398 (1996).
18. Shao, Z., and Burkhalter, A., *J. Neurosci.* **16**, 7353-7365 (1996).
19. Moore, T., and Fallah, M., *J. Neurophysiol.* **91**, 152-162 (2004).
20. Gotts, S. J., Gregoriou, G. G., Zhou, H., and Desimone, R. "Synchronous Activity Within And Between Areas V4 And FEF In Attention", *Soc. Neurosci. Abs. 703.7,* 2006.
21. Borgers, C., Epstein, S., and Kopell, N. J., *Proc. Nat. Acad. Sci. USA* **102**, 7002-7007 (2005).
22. Deco, G., and Rolls, E. T., *J. Neurophysiol.* **94**, 295-313 (2005).
23. Buia, C., and Tiesinga, P., *J. Comput. Neurosci.* **20**, 247-264 (2006).
24. Ardid, S., Wang, X. J., and Compte, A., *J. Neurosci.* **27**, 8486-8495 (2007).

AUTHOR INDEX

A

Aguilera, P., 98
Alvares-Ramirez, J., 87

B

Barrera, D., 98
Berezkovskii, A. M., 11
Breña-Medina, V. F., 115

C

Chánez-Cárdenas, M. E., 98

D

Dagdug, L., 11, 57

E

Echeverría, J. C., 87
Estrada-Gil, J. K., 34

F

Fernández-López, J. C., 34

G

García-Colín Scherer, L.S., 109

H

Hernández-Lemus, E., 34
Hidalgo-Miranda, A., 34
Hugues, E., 135

J

Jiménez-Sánchez, G., 34
José, J. V., 135

L

Lerma, C., 87

M

Maldonado, P. D., 98

O

Olivares-Quiroz, L., 109
Ortiz-Plata, A., 98

R

Rodríguez, E., 87

S

Sánchez-Garduño, F., 115
Seol, Y., 1
Silva-Zolezzi, I., 34
Skinner, G. M., 1